T0344862

Geothermal Energy

Geothermal Energy

Sustainable Heating and Cooling Using the Ground

Marc A. Rosen and Seama Koohi-Fayegh

University of Ontario Institute of Technology, Oshawa, Canada

This edition first published 2017

© 2017 John Wiley & Sons, Ltd

Registered Office
John Wiley & Sons Ltd, The Atrium, Southern Gate, Chichester, West Sussex, PO19 8SQ, United Kingdom

For details of our global editorial offices, for customer services and for information about how to apply for permission to reuse the copyright material in this book please see our website at www.wiley.com.

Library of Congress Cataloging-in-Publication Data

Names: Rosen, Marc (Marc A.), author. | Koohi-Fayegh, Seama, 1983- author.
Title: Geothermal energy : sustainable heating and cooling using the ground /
 Marc A. Rosen and Seama Koohi-Fayegh, University of Ontario Institute of
 Technology, Oshawa, Canada.
Description: Chichester, West Sussex, United Kingdom : John Wiley & Sons,
 Inc., [2017] | Includes bibliographical references and index.
Identifiers: LCCN 2016037031 (print) | LCCN 2016046782 (ebook) | ISBN
 9781119180982 (cloth) | ISBN 9781119181033 (pdf) | ISBN 9781119181019
 (epub)
Subjects: LCSH: Ground source heat pump systems.
Classification: LCC TH7417.5 .R67 2017 (print) | LCC TH7417.5 (ebook) | DDC
 697/.7–dc23
LC record available at https://lccn.loc.gov/2016037031

A catalogue record for this book is available from the British Library.

Set in 10/12pt Warnock by SPi Global, Chennai, India

Printed and bound in Malaysia by Vivar Printing Sdn Bhd

10 9 8 7 6 5 4 3 2 1

To my wife, Margot, and my children, Allison and Cassandra, for their inspiration and love.

And to my parents for their love and support.

Marc A. Rosen

To my husband, Ali, and my mother, for their inspiration and love.

And to the memory of my father.

Seama Koohi-Fayegh

Contents

Preface

Geothermal energy systems that provide heating and cooling using the ground are increasingly applied, and represent a technology that supports sustainable use of energy. Ground-source heat pumps, thermal energy storage and district energy are components of geothermal energy systems, and have been around for over 40 years and are widely applied. But they are also undergoing research and being improved continually, and advanced systems and components, as well as advanced understanding, are expected to be developed over the foreseeable future.

In this book, geothermal energy systems that utilize ground energy in conjunction with heat pumps to provide sustainable heating and cooling are described. Information on a range of topics is provided, from thermodynamic concepts to more advanced discussions on the renewability and sustainability of closed-loop geothermal energy systems. Numerous applications of such systems are also described. Theory and analysis are emphasized throughout, with detailed descriptions of models available for vertical geothermal heat exchangers.

The book also contains many references, including some related to books and articles on various aspects of geothermal systems that are not fully covered. Some links to websites with basic freeware for ground-source heat transfer modeling and building heating loads are referenced throughout the book.

The book is research oriented, thereby ensuring that new developments and advances in geothermal energy systems are covered.

The book is intended for use by advanced undergraduate or graduate students in several engineering disciplines such as mechanical engineering, chemical engineering, energy engineering, environmental engineering, process engineering and industrial engineering. Courses on geothermal energy systems or related courses such as heat exchangers, thermal energy storage or heat pumps that are often offered at the graduate level in Mechanical Engineering or related fields may find this book useful. The information included is sufficient for energy, environment and sustainable development courses. The book can also be used in research centers, institutes and labs focusing on the areas mentioned above, by related learned societies and professional associations, and in industrial organizations and companies interested in geothermal energy and its applications. Drillers and installers as well as regulatory agencies may also be interested in the book. Furthermore, the book offers a valuable and readable reference text source for anyone interested in learning about geothermal energy systems.

The book strives to provide clear information on ground-based geothermal systems and the many advances occurring in the field in a way that makes it understandable for students, practitioners, researchers and policy makers.

Various topics are covered, from fundamentals to advanced discussions on sustainability. Many applications are described, while theory and analysis are emphasized throughout. Detailed descriptions are provided of models for geothermal heat exchangers and heat pumps. The organization of the book is intended to help the reader build knowledge in a logical fashion while working through the book, and is as outlined here. Introductory material is included in the first two chapters, with an overview of geothermal energy as a source of energy and technologies that can harvest it described in Chapter 1, and fundamentals of thermofluid engineering disciplines related to geothermal energy systems provided in Chapter 2. Information on the main components of geothermal energy systems such as heat pumps, heat exchangers, heating, ventilating, and air conditioning equipment and energy storage units are provided in Chapter 3. The next five chapters form the heart of the book, with thermal energy storage being the focus of Chapter 4, geothermal heating and cooling forming the core of Chapter 5, and design and installation considerations for geothermal energy systems being the emphasis of Chapter 6. Extensive material is provided on modeling of ground heat exchangers and heat pumps, with the modeling of ground heat exchangers including a variety of models examined in Chapter 7 and the application of the models to various relevant examples presented in Chapter 8. The thermodynamic analysis of geothermal energy systems is the focus of Chapter 9. Extensive coverage is provided on environmental and sustainability factors, as these have become increasingly germane in recent years. Environmental factors related to geothermal energy systems are covered in Chapter 10 while their renewability and sustainability are examined in Chapter 11. To close, a range of case studies for geothermal energy systems is presented in Chapter 12 that illustrate the technologies, their applications and their advantages and disadvantages.

The main features of the book are:

- comprehensive coverage of ground-based geothermal energy systems;
- detailed descriptions and discussions of methods to determine potential environmental impacts of geothermal energy systems and their thermal interactions;
- presentations of the most up-to-date information in the area;
- suitability as a good reference for geothermal heat exchangers;
- a research orientation to provide coverage of the state of the art and emerging trends and recent developments;
- numerous illustrative examples and case studies;
- clarity and simplicity of presentation of geothermal energy systems that use the ground.

We hope this book allows geothermal energy to be used more widely for the provision of heating and cooling services using the ground in a sustainable manner, using both existing and conventional equipment and systems as well as new and advanced technologies. The book aims to provide an enhanced understanding of the behaviours of heating and cooling systems in the form of ground-source heat pumps that exploit geothermal energy for sustainable heating and cooling of buildings, and enhanced tools for improving them. By exploiting the benefits of applying exergy methods to these ground-based energy systems, we believe they can be made more efficient, clean and sustainable, and help humanity address many of the challenges it faces.

October 2016

Marc A. Rosen and Seama Koohi-Fayegh
University of Ontario Institute of
Technology, Oshawa, Canada

About the Authors

Marc A. Rosen is a Professor at the University of Ontario Institute of Technology in Oshawa, Canada, where he served as founding Dean of the Faculty of Engineering and Applied Science. A former President of the Engineering Institute of Canada and the Canadian Society for Mechanical Engineering, he is a registered Professional Engineer in Ontario. He has served in many professional capacities, including Editor-in-Chief of several journals and a member of the Board of Directors of Oshawa Power and Utilities Corporation. He is an active teacher and researcher in energy, sustainability, geothermal energy and environmental impact. Much of his research has been carried out for industry, and he has written numerous books. He has worked for such organizations as Imatra Power Company in Finland, Argonne National Laboratory near Chicago, and the Institute for Hydrogen Systems near Toronto. He has received numerous awards and honours, including an Award of Excellence in Research and Technology Development from the Ontario Ministry of Environment and Energy, the Engineering Institute of Canada's Smith Medal for achievement in the development of Canada, and the Canadian Society for Mechanical Engineering's Angus Medal for outstanding contributions to the management and practice of mechanical engineering. He is a Fellow of the Engineering Institute of Canada, the Canadian Academy of Engineering, the Canadian Society for Mechanical Engineering, the American Society of Mechanical Engineers, the International Energy Foundation and the Canadian Society for Senior Engineers.

Seama Koohi-Fayegh is a Post-doctoral Fellow at the Department of Mechanical Engineering at the University of Ontario Institute of Technology in Oshawa, Canada. She received her PhD in Mechanical Engineering at the University of Ontario Institute of Technology under the supervision of Professor Marc A. Rosen. Her PhD thesis topic was proposed by the Ontario Ministry of Environment and focused on thermal sustainability of geothermal energy systems: system interactions and environmental impacts. She did her Master's degree in Mechanical Engineering (Energy Conversion) at Ferdowsi University of Mashhad, Iran, and worked on entropy generation analysis of condensation with shear stress on the condensate layer. Her thesis research won multiple awards at the school level and at the Iranian Society of Mechanical Engineering in 2009. Her research interests include heat transfer, sustainable energy systems and energy technology assessment.

About the Authors

Acknowledgments

The work of many of our colleagues helped greatly in the development of this book, and is gratefully acknowledged. Some of the material in this book is derived from research that we have carried out with numerous distinguished collaborators over the years. These include the following faculty members in geothermal energy and related areas:

- Drs Ibrahim Dincer and Bale V. Reddy, University of Ontario Institute of Technology, Oshawa, Ontario, Canada
- Drs Wei Leong and Alan Fung, Ryerson University, Toronto, Ontario, Canada
- Dr Vlodek R. Tarnawski, Saint Mary's University, Halifax, Nova Scotia, Canada
- Dr Robert A. Schincariol, Western University, London, Ontario, Canada
- Dr Tomasz Śliwa, AGH University of Science and Technology, Krakow, Poland
- Dr Frank C. Hooper, University of Toronto, Toronto, Ontario, Canada
- Dr David S. Scott, University of Victoria, Victoria, British Columbia, Canada

We highly appreciate all of their efforts, as well as their thought-provoking insights.

Last but not least, the authors warmly thank their families, for their endless encouragement and support throughout the completion of this book. Their patience and understanding is most appreciated.

Nomenclature

A	surface area; m^2; cross-sectional area, m^2
a	absorptivity; temperature coefficient; constant
b	constant
Bi	Biot number
C	volumetric heat capacity of soil, J/m^3 K
c	specific heat, J/kg K
COP	coefficient of performance
c_p	specific heat at constant pressure, J/kg K
c_v	specific heat at constant volume, J/kg K
d	diameter, m
dA	surface element, m^2
d_i	pipe inner diameter, m
D	pipe diameter, m; uppermost part of the borehole, m; U-tube leg half distance, m
D_b	borehole separation distance, m
D_ϑ	isothermal moisture diffusivity, m^2/s
E	energy, kJ; electrical energy, kJ
e	specific energy, J/kg
Ex	exergy, kJ
ex	specific exergy (flow or non-flow), kJ/kg
Ex^Q	exergy transfer associated with heat transfer, kJ
$\dot{Ex}$	exergy rate, kW
Ex_{dest}	exergy destruction (irreversibility), kJ
$\dot{Ex}_{\text{dest}}$	exergy destruction rate (irreversibility rate), kW
F	force, N
f	fraction
Fo	Fourier number
Gr	Grashof number
Gz	Graetz number
g	gravitational acceleration, m/s^2
H	overall heat transfer coefficient, W/m^2 K; active borehole length, m
$\overline{H}$	dimensionless parameter [Equation (7.28)]
h	specific enthalpy, kJ/kg; heat transfer coefficient, W/m^2 K; borehole distance from coordinate center, m
h_i	heat transfer coefficient of circulating fluid, W/m^2 K

h_z	depth where borehole heating starts, m
J_0	Bessel function of the first kind, order 0
J_1	Bessel function of the first kind, order 1
K_h	hydraulic conductivity, m/s
k	adiabatic exponent; thermal conductivity, W/m K; ground thermal conductivity, W/m K
k_b	grout thermal conductivity, W/m K
L	length scale, m; latent heat of vaporization of water, J/kg
L_s	leg spacing of U-tube, m
l	position of borehole in x coordinate, m
m	mass, kg
$\dot{m}$	mass flow rate, kg/s; borehole fluid mass flow rate, kg/s
M	molecular weight, kg/mol
n	number of moles, mol; number of time steps
NTU	number of transfer units
Nu	Nusselt number
P, p	pressure, Pa; dimensionless parameter [Equation (7.22)]
Pe	Peclet number
Pr	Prandtl number
Q	heat transfer, J
$\dot{Q}$	heat transfer rate, W
q	thermal radiation rate, W
q'	heat flow rate per unit length of borehole, W/m
q'_1	heat flow rate per unit length of inlet pipe, W/m
q'_2	heat flow rate per unit length of outlet pipe, W/m
q''	heat flux at borehole wall, W/m^2
$\dot{q}_{gen}$	generated heat per unit volume, W/m^3
R	gas constant, J/kg K; thermal resistance, m K/W
$\overline{R}$	universal gas constant, 8.314 kJ/mol K; dimensionless parameter [Equation (7.29)]
Ra	Rayleigh number
Re	Reynolds number
R_{11}	thermal resistance between the inlet pipe and the borehole wall, m K/W
R_{12}	thermal resistance between the inlet and outlet pipes, m K/W
R_{22}	thermal resistance between the outlet pipe and the borehole wall, m K/W
R_1^Δ	thermal resistance between Pipe 1 and borehole wall, m K/W [Equation (7.18)]
R_2^Δ	thermal resistance between Pipe 2 and borehole wall, m K/W [Equation (7.18)]
R_{12}^Δ	thermal resistance between Pipes 1 and 2, m K/W [Equation (7.18)]
R_{b2}	thermal resistance between circulating fluid and borehole wall based on two-dimensional analysis, m K/W
R_{b3}	thermal resistance between circulating fluid and borehole wall based on three-dimensional analysis, m K/W
R_g	thermal resistance of conduction in grout, m K/W
R_p	thermal resistance of conduction in pipe, m K/W

R_s	ratio of ground heat extraction to ground heat injection
r	reflectivity; radial scale, m; radial coordinate, m
r^*	direction perpendicular to U-tube surface
r^{**}	radial distance from borehole axis
r_1	distance of point (x,y) in soil around multiple boreholes from Borehole 1, m
r_2	distance of point (x,y) in soil around multiple boreholes from Borehole 2, m
r_b	borehole radius, m
r_i	pipe inner radius, m
r_p	pipe radius, m
S	entropy, kJ/K
s	specific entropy, kJ/kg K
St	Stanton number
S_ϕ	source of ϕ per unit volume
T	temperature, K or °C
T'_f	temperature of borehole fluid entering U-tube, K
T''_f	temperature of borehole fluid exiting U-tube, K
t	transmissivity; time, s
t_s	steady-state time, s
Δt	time step, s
U, u	velocity, m/s
u	specific internal energy, kJ/kg; velocity in x direction, m/s
V	volume, m³; velocity, m/s
$\dot{V}$	volumetric flow rate, m³/s
v	specific volume, m³/kg; kinematic viscosity, m²/s; velocity in y direction, m/s; velocity, m/s
$\bar{v}$	molar volume, m³/mol
v_0	velocity of borehole fluid, m/s
W	shaft work, J
$\dot{W}$	work rate or power, kW
w	velocity in z direction, m/s; position of borehole in y coordinate, m
x	x coordinate, m
Y	characteristic length, m
Y_0	Bessel function of the second kind, order 0
Y_1	Bessel function of the second kind, order 1
y	y coordinate, m
Z	dimensionless depth [Equation (7.22)]
z	axial coordinate, m; depth, m
∇	differential operator, del

Greek Letters

α	thermal diffusivity, m²/s
β	dimensionless parameter [Equation (7.22)]
β_0	shape factor of grout resistance [Equation (7.10)]

β_1	shape factor of grout resistance [Equation (7.10)]
Γ_ϕ	diffusion coefficient for ϕ
γ	Euler's constant, 0.5772
$\triangle$	difference
$\Delta\xi$	distance between centroids A and P of two neighboring grids, m
ε	emissivity; heat exchanger effectiveness; heat transfer efficiency of borehole; phase conversion factor
η	energy efficiency
Θ	dimensionless temperature [Equation (7.22)]
ϑ	volumetric moisture content (dimensionless); temperature difference relative to ground initial temperature, K; parameter
ϑ_l	volumetric liquid content (dimensionless)
μ	chemical potential, J/mol; dynamic viscosity, N s/m^2
ρ	density, kg/m^3
σ	Stefan–Boltzmann constant, 5.669×10^{-8} W/m^2 K^4
τ_i	time at which step heat flux q_i is applied, s
Φ	scalar quantity
φ	circumferential coordinate, rad
ψ	exergy efficiency

Subscripts

0	initial, ambient or reference condition
A	centroid A
a	surroundings
adv	advective
ave	average
b	borehole
bal	balance
BHE	borehole heat exchanger
BW	BHE side water/glycol solution
c	cell; cooling; charging
CL	cooling load
comp	compressor
cond	condenser
CS	control surface
CV	control volume
CW	cooling water
d	discharging
dest	destruction
e	exit; evaporation, evapotranspiration, melting snow or sublimation; equivalent
evap	evaporator
ExpV	expansion valve
f	fluid; borehole fluid; final
$f1$	borehole fluid in inlet pipe

$f2$	borehole fluid in outlet pipe
FanCoil	fan coil
g	grout; ground
H	high-temperature; heating; high
h	convective; heating
HL	heating load
i	initial; inlet; inner; ith borehole; ground discretization designation in r direction; ith time step
in	inlet
L	low-temperature; low
L	liquid water
lo	long-wave radiation
max	maximum
n	normal
nb	node number of adjacent cell
o	outdoor; overall; reference-environment state
out	outlet
P	centroid P
p	pipe
pump	pump
PHX	plate heat exchanger
r	radiation; in radial direction
rev	reversible
s	surface; ground
sn	shortwave radiation
sys	system
t	threshold
valve	valve
z	in axial direction
ϑ	in circumferential direction

Superscripts

.	rate with respect to time
0	previous time step
n	polytropic exponent; discretization step designation in time

Abbreviations

ASHP	air-source heat pump
ASHRAE	American Society for Heating, Refrigerating and Air-conditioning Engineers
BHE	borehole heat exchanger
BTES	borehole thermal energy storage
DLSC	Drake Landing Solar Community

GHE	ground heat exchanger
GHG	greenhouse gas
GSHP	ground-source heat pump
HGHE	horizontal ground heat exchanger
HP	heat pump
HVAC	heating, ventilating, and air conditioning
IEA	International Energy Agency
O	order
PCM	phase-change material
SUP	supplementary
TES	thermal energy storage
TRT	thermal response test
VGHE	vertical ground heat exchanger
ε-NTU	effectiveness-number of transfer units

1

Introduction to Geothermal Energy

Geothermal energy systems are one option for providing energy services. They take advantage of the ground and the energy it contains. Sometimes ground energy is the basic ground at its natural temperature, which is mainly affected by ambient conditions. At other times, the ground is at an elevated temperature. Considering the current level of geothermal energy use and future energy needs, geothermal energy sources show great potential for contributing a larger fraction of the world's energy needs.

Archaeological evidence shows that geothermal energy was first used by ancient peoples, including the Romans, Chinese, and Native Americans. They used hot mineral springs as a source of heat for bathing, cooking, and heating. The minerals in water from these springs also served as a source of healing. While such uses of hot springs have changed over time, they are still used as a source of heat for bathing in several spas around the world. With technological developments, the use of geothermal energy has expanded to deeper levels of the earth's crust, which can be used for a wider range of applications such as domestic heating and cooling, industrial processes, and electricity generation. However, only a small fraction of available geothermal energy is currently used commercially to generate electricity or provide useful heating, in part due to the current state of the technology.

Geothermal energy systems that exploit hot reservoirs in the ground (e.g., thermal springs, geysers, ground heated by hot magma) are used mainly to generate electricity and to provide heating. Such systems are common in countries such as Iceland, Turkey and others. The global operating capacity for geothermal electricity generation from such geothermal resources is about 12.8 GW as of January 2015, spread across 24 countries, and it is expected to reach between 14.5 GW and 17.6 GW by 2020 (Geothermal Energy Association 2015).

There is another type of geothermal energy system, which provides heating and cooling using the ground. That is the type of geothermal energy that is the focus of this book. Such geothermal energy systems take advantage of the energy contained in the ground in its natural state, even when it is not at elevated temperatures due to heat within the earth. This ground energy is related to the background ground temperature and includes the ground itself and groundwater.

Geothermal Energy: Sustainable Heating and Cooling Using the Ground, First Edition.
Marc A. Rosen and Seama Koohi-Fayegh.
© 2017 John Wiley & Sons, Ltd. Published 2017 by John Wiley & Sons, Ltd.

1.1 Features of Geothermal Energy

Ground-based energy can be used in all seasons:

- Ground-based energy can provide heating directly in winter, since the ground below the surface is often warmer than the air above. Such applications include space heating, greenhouse heating, aquaculture pond heating, agricultural drying, industrial heating uses, bathing and swimming, and snow melting. Sometimes the ground temperature is only adequate to provide preheating. The ground temperature can also be boosted via devices like heat pumps, allowing ground-based energy to provide heating at higher temperatures. The use of geothermal energy via ground-source heat pumps has grown considerably compared to the other applications, primarily due to the technology's ability to achieve high efficiency and to utilize groundwater and/or ground temperature anywhere in the world.
- Conversely, ground-based energy can provide direct cooling in summer, since the ground below the surface is often cooler than the hot air above. Again, the ground temperature may only be adequate to provide precooling. But the ground temperature can also be lowered using heat pumps operating in a cooling mode, allowing ground-based energy to provide cooling at lower temperatures.

Although the earth's ultimate geothermal energy potential cannot be estimated based on our current level of knowledge and the unpredictability of technology development, geothermal energy systems of both types are usually classified as renewable energy forms. When such geothermal energy is utilized, the temperature of the ground is returned to its elevated temperature by heat contained within hot regions in the earth, or by the effect of the ambient conditions. Discussions of the renewability of various heat sources vary for the different technologies utilizing the energy source. For example, technologies that utilize the ground at temperatures affected by the ambient conditions can be considered renewable provided the ambient conditions are sustained. The constant heat supply from solar radiation and the sustainability of the hydrological cycle (infiltration and precipitation) guarantees a constant flow of heat to the ground and the renewability of such geothermal sources. The energy replacement often occurs on a time scale comparable with that of the extraction time scale.

Sustainable geothermal energy utilization often refers to how this energy resource is used to meet current energy needs without compromising its future utilization. Estimating the long-term response of geothermal energy sources to current utilization and production capacity levels is important if we are to understand their potential contributions to sustainable development. As a renewable energy source, geothermal energy is often viewed as a contributor to sustainable development and the broader goal of sustainability, provided that they are well designed. Being sustainable goes beyond geothermal energy being a renewable energy form, and includes many of its other characteristics:

- **Availability.** Geothermal energy in the form of ground at elevated temperature is available in many parts of the world, especially in regions with seismic and volcanic activity. Geothermal energy in the form of ground at ambient temperature is available almost everywhere, although its temperature depends on the location and climate. Geothermal energy is available day and night, every day of the year, and can thus cover base-load energy needs and serve as a supplement to intermittent energy sources. The availability characteristics of intermittent renewable energy forms such as solar and wind are much different.

- **Compatibility.** Systems exploiting geothermal energy are often compatible with both centralized and distributed energy generation.
- **Affordability.** Geothermal energy is often exploitable for heating and cooling, and for electricity generation, in an affordable manner. Of course, some geothermal systems are not economically viable, but work is ongoing on several of these to improve commercial prospects.
- **Acceptability.** Most people are supportive of geothermal energy, in part because it is renewable and often economically viable, and also because geothermal energy systems are not intrusive and usually are invisible. This is not the case for many other renewable energy forms, such as solar and wind.

Barriers to deployment include high capital costs, resource development risks, lack of awareness about geothermal energy, and perceived or real environmental issues.

1.2 Geothermal Energy Systems

Geothermal energy systems can exploit hot reservoirs in the ground, often in the form of natural hot water or steam, to provide heating and electricity generation. The geothermal energy technologies that are used in electricity generation are flash technologies, including double and triple flash units, dry steam, and binary cycles. Electricity generation using flash technologies contribute to nearly 60% of the global market use, with dry steam and binary cycles accounting for 26 and 15% of the global market, respectively (Geothermal Energy Association 2015). Growth in use of such geothermal energy systems for heating and electricity generation is limited by their high capital costs. Geothermal development costs depend on resource temperature and pressure, reservoir depth and permeability, fluid chemistry, location, drilling requirements, size of development, number and type of plants (dry steam, flash, binary or hybrid) used, and whether the project is greenfield or expansion (10–15% less). Development costs are strongly affected by prices of commodities (e.g., oil, steel, and cement). Declines in oil and gas prices can decrease geothermal capital costs.

Geothermal energy systems that provide heating and cooling using the ambient ground are made up of various systems and components. Some of the main systems include ground-source heat pumps, thermal energy storage systems, and district energy (i.e., district heating and/or district cooling) capabilities. They also include many other components, such as compressors, heat exchangers, pumps and pipes. Ground-source heat pump systems are capable of providing heating and cooling in one unit. The capacity of a ground-source heat pump is selected based on the heating and cooling loads, the temperature of the ground, and other parameters. Since most areas do not have balanced heating and cooling loads, the capacity of the heat pump is often selected based on one load. In most regions in the USA, the heat pump capacity matches the cooling load and is oversized for the heating loads. In Europe, ground-source heat pumps are used in the residential sector to cover base heating loads and are integrated with another heating system that covers peak heating loads. The capacity of individual ground-source heat pump units ranges from about 1.5 t for small residential applications to over 40 t for commercial and institutional applications. Technology improvements in ground-source heat pumps are expected to improve the performance and lower the cost of heat pump technologies. Key components such as compressors and heat exchangers will likely provide the largest areas for improvement. The main goals for ground-source heat pumps are reducing capital costs and improving operating efficiency, while expanding the range of products for most of the heating and cooling applications and sub-markets in the building sector.

Thus, geothermal energy systems can provide heating and cooling using the ambient ground, and can exploit hot reservoirs in the ground to provide heating and electricity generation. Both types of geothermal energy are used in practise, and are finding increased application. But the use of geothermal energy systems that use the ground to provide heating and cooling services (the focus of this book) is growing at a particularly noteworthy rate. According to a recent report (Lund and Boyd 2015), the direct use of geothermal energy has experienced an annual growth of 7.7% in capacity over the 5-year period after 2010, with the highest installed thermal capacity in the USA, China, and Sweden. This growth is mainly attributable to the growing popularity of ground-source heat pumps. About 90 000 TJ/year of ground-source heat pump utilization was observed in 2010, and this grew to approximately 325 000 TJ/year by the end of 2014.

Although geothermal energy technologies have been around for over 40 years and are applied in many areas, they are continually undergoing research and development. These efforts allow for system improvements, advances in components and enhanced understanding. Such activity is likely to carry on in the future.

1.3 Outline of the Book

In this book, geothermal energy systems are described that utilize ground energy in conjunction with heat pumps to provide heating and cooling, in a sustainable fashion. Various topics are covered, from thermodynamic fundamentals to advanced discussions on renewability and sustainability. Many applications of such systems are also described, while theory and analysis are emphasized throughout. Detailed descriptions are provided of models for vertical geothermal heat exchangers, and a strong focus is placed on closed-loop geothermal energy systems.

In this chapter, an introduction to geothermal energy as a source of energy and technologies that can harvest it is provided. Some key features of geothermal energy systems, such as its renewability and sustainability, as well as some of its advantages are briefly described. The main components of such systems are reviewed. The aim of this chapter is to provide the reader with basic information to help develop an understanding of the overall scope and range of material that is included in this book.

In Chapter 2, fundamentals of thermodynamics, heat transfer and fluid mechanics that are related to geothermal energy systems are provided to familiarize readers with these topics and prepare them for subsequent chapters. A good knowledge of thermodynamics is important to understanding geothermal energy, especially heat pumps. Facets of thermodynamics most relevant to geothermal energy systems and their applications are introduced and particular attention is paid to the quantity exergy and the methodology derived from it, exergy analysis. Aspects of heat transfer relevant to geothermal energy systems are introduced to provide the reader with a good grounding in heat transfer, which is central to geothermal energy utilization and its application.

The three main modes of heat transfer are considered: conduction, convection, and radiation. A good grounding of fluid mechanics helps in understanding geothermal energy systems, as fluid flow problems often arise, so elements of fluid mechanics relevant to geothermal energy systems are also introduced. Finally, basic concepts about the ground are presented, since such material is fundamental to understanding ground-based geothermal systems, including information on ground temperature range and gradients, ground properties, and the existence of ground-based ecosystems and their sensitivity to human activity in the ground.

Chapter 3 provides background information on components of geothermal energy systems such as heat pumps, heat exchangers, heating, ventilating and air condition-ing (HVAC) equipment and energy storage units. An understanding of these technolo-gies is needed to analyze them and the larger energy systems that they comprise. The devices examined include heat pumps, which are cyclic devices that transfer heat from a low-temperature medium to a high-temperature medium; heat exchangers, which are devices for heat exchange processes between two media; HVAC equipment, which provides heating, cooling, humidification and dehumidification for spaces, and energy storage systems that permit harvested or otherwise available energy to be stored until such a time when it is needed or desired.

Chapter 4 focuses on thermal energy storage (TES) concepts, theory and applications. Details on thermal storage types, operation and applications are provided, for both heat and cold storage. The main thermal storage types, sensible, latent and thermochemical, are covered. A focus is placed on underground thermal energy storages, which normally are sensible storages, as they can store both hot and cold energy in the ground and thus are often integral to geothermal energy systems. Common types of underground thermal energy storage are described: soil and earth bed, borehole, aquifer, rock cavern, container/tank, and solar pond. Finally, the integration of thermal energy storage with heat pumps is examined, as such systems can be particularly beneficial for heating and cooling applications.

Geothermal heating and cooling is the focus of Chapter 5. Ground-based energy can provide heating in winter and cooling in summer, in partial or full manners. Ground-source heat pumps normally form the basis of such systems and therefore are extensively discussed. Emphasis is placed on geothermal heat exchangers (also called ground, underground and ground-coupled heat exchangers), since they facilitate the exchange of heat between a fluid and the ground. For completeness, high-temperature geothermal systems are also described, including systems that essentially use the ground as a heat source for electricity generation and heating.

General information on design considerations for geothermal energy systems and procedures for the installation of ground-source heat pump systems are provided in Chapter 6. The material on how systems are designed describes the relation to building loads and weather. Procedures in designing systems with unbalanced loads are also dis-cussed. Building energy calculations, which are important first steps in designing any space heating and cooling system, are covered, as are building and heat pump perfor-mance considerations, heating and cooling calculations, and ground heat injection and extraction. Finally, the economics of geothermal systems, which vary according to type and application, is described.

The modeling of ground heat exchangers is examined in Chapter 7. Various models have been reported for heat transfer in borehole heat exchangers and their coupling with HVAC and building energy systems. Both analytical and numerical approaches are considered, and parameters such as moisture migration and groundwater flow, relevant boundary conditions, and solution errors are described. Groundwater and how it and its movement affects the performance of ground-source heat pump systems is assessed. Two- and three-dimensional models are covered, as are horizontal and vertical heat exchangers. Finally, the coverage is expanded to the modeling of multiple boreholes in the form of a borefield and systems with time varying heat transfer rates.

In Chapter 8, the analytical and numerical models presented in Chapter 7 are applied to various examples, involving various levels of difficulty and detail. Since the modeling approach is often selected based on the objective of modeling ground heat exchangers, the objectives of modeling the systems are described for each example, and challenges of each model example are explained. The examples include the semi-analytical modeling of two boreholes, the numerical modeling of two boreholes, the numerical modeling of a borefield, and the numerical modeling a horizontal ground heat exchanger.

The thermodynamic analysis of geothermal energy systems of various kinds is the focus of Chapter 9. Special attention is placed on the application of exergy concepts in analysis of energy systems to evaluate and optimize their design and performance. The energy systems discussed include an underground thermal storage in the form of an aquifer TES, a hybrid ground-source heat pump system, and ground-source heat pumps and underground thermal storage. A thermodynamic analysis of a complex system integrating ground-source heat pumps and underground thermal storage is featured.

Environmental factors of relevance to geothermal energy systems are covered in Chapter 10. These include reductions in greenhouse gas emissions achievable with such systems. From an environmental perspective, preserving natural habitats in the ground by avoiding drastic changes in temperature and moisture content is an important subject especially when discussing the installation of ground heat exchangers (GHEs) at various depths in the ground. Geothermal system installation can incur environmental benefits and impacts, so both are discussed. The "thermal pollution" released from ground-source heat pump systems to the ground and potentially sensitive ecosystems is described. The chapter includes material that can guide regulatory agencies and industry towards designs and installations that improve and even maximize their sustainability and reduce or minimize possible environmental impacts.

Chapter 11 covers the renewability and sustainability of geothermal energy systems. As a renewable energy source, geothermal energy is often seen to have a significant role to play as a contributor to sustainable development and, more broadly, sustainability. Particular attention is paid to the thermal interactions between ground-source heat pumps and their underground heat exchangers.

In Chapter 12, a range of case studies is presented to illustrate the application of geothermal energy systems that utilize the ground for heating and cooling as well as their advantages and disadvantages. The cases consider applications from the residential, commercial and institutional building sectors, as well as relevant utility sector entities involved in electricity generation and district heating and cooling. The case studies illustrate the context in which a geothermal energy system can be employed and assessed, and are based mainly on actual applications and drawn from various sources. The types of geothermal energy systems covered through the case studies include an underground thermal energy storage, a ground and water tank thermal energy storage for heating, a space conditioning with a heat pump and seasonal thermal storage, and an integrated system with a ground-source heat pump, thermal storage and district energy.

For further reading, many references on various aspects of geothermal systems are given, and links to websites with basic freeware for ground-source heat transfer modeling and building heating loads are provided.

References

Geothermal Energy Association (2015) Annual U.S. & Global Geothermal Power Production Report', http://geo-energy.org/reports.aspx (accessed March 9, 2016).

Lund, J.W. and Boyd, T.L. (2015) Direct utilization of geothermal energy 2015 worldwide review. *Proceedings of the World Geothermal Congress*, April 19–25, 2015, Melbourne, Australia. International Geothermal Association.

2

Fundamentals

2.1 Introduction

In this chapter, the fundamentals of thermodynamics, heat transfer and fluid mechanics that are related to heat pumps and heat exchangers are provided to familiarize the reader with the basis of the analyses in subsequent chapters. Furthermore, a section on nature of the ground provides information on ground temperature ranges and gradients, ground properties, the existence of ecosystems, and their sensitivities to human activity in the ground.

2.2 Thermodynamics

A good knowledge of thermodynamics is important to understanding geothermal energy. Facets of thermodynamics most relevant to geothermal energy systems and their applications are introduced in this section to provide the reader with a solid grounding in the fundamentals of thermodynamics.

2.2.1 Thermodynamic System, Process and Cycle

A thermodynamic system is a region, device or group of devices that contains a quantity of matter. A system is delineated by its boundaries.

Depending on whether or not matter crosses a system's boundaries, two system types can be defined:

- **Open.** A system can in general experience material flows across its boundaries (Figure 2.1).
- **Closed.** A closed system experiences no material flows across its boundaries and thus contains a fixed quantity of matter (Figure 2.2). Note that a closed system, for which no mass, heat or work interactions cross its boundaries, and which is therefore not affected by the surroundings, is referred to as an isolated system.

A process is the conversion of energy from one form to another, or a physical or chemical change in the properties of matter. One material property remains constant for some processes. Examples include constant temperature (isothermal), constant pressure (isobaric) and constant volume (isochoric) processes.

A cycle is a series of thermodynamic processes for which the material properties are the same at the start and end.

Figure 2.1 An open system.

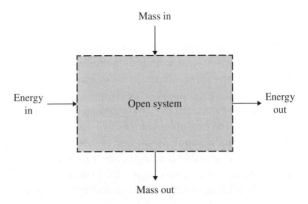

Figure 2.2 A closed system.

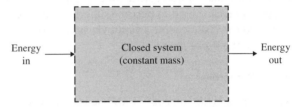

2.2.2 Thermodynamic Property

A thermodynamic property is a characteristic of a quantity of matter, for example, internal energy, enthalpy, entropy, temperature, pressure and density, used to describe its thermodynamic state.

There are two main types of thermodynamic properties:

- **Intensive.** Intensive properties are independent of size or scale. Examples include temperature, pressure, and density.
- **Extensive.** Extensive properties are dependent on size or scale, and thus change as the size of a system varies. Examples include mass, volume, internal energy, and entropy. Note that extensive properties expressed per unit mass are intensive properties (e.g., specific volume).

The state of a substance can usually be specified if two independent thermodynamic properties of the matter are known. Once the state is specified, all other properties can be determined.

A substance may be pure or not. A pure substance has a homogeneous and constant chemical composition. A pure substance may be in one phase, for example, liquid water, or more than one phase, for example, a mixture of ice and liquid water or a mixture of liquid water and water vapor. Each phase has the same chemical composition. A substance is not pure if its parts have differing chemical compositions. For instance, a mixture of two different liquids or two different gases is not a pure substance.

2.2.3 State and Phase

The state of a system or substance is defined as its condition, as characterized by such properties as temperature and pressure. The properties of a substance in a given state each have a single value. State properties are independent of process path, that is, the processes the substance underwent to reach the state.

When energy is added to or removed from a substance, it undergoes a change of state. Sometimes when heat is added or removed, the substance undergoes a change of state but stays in the same phase. In other instances, heat addition or removal also causes a phase change (e.g., liquid to or from solid, vapor to or from liquid). The temperature remains constant until the phase change is complete, for a given pressure.

Consider, for example, the case of ice melting to water, followed by water boiling and becoming a vapor. If carried out isobarically, the water undergoes the following processes: (1) sensible heating of the liquid water from its initial temperature to the saturation temperature, making it a fully saturated liquid (i.e., water with a vapor quality of zero); (2) constant-temperature vaporization, converting the saturated liquid to a saturated vapor, that is, water with a vapor quality of 1; (3) sensible heating of the saturated water vapor, causing it to become superheated as its temperature rises.

A solid can also become a vapor via sublimation, a process by which the solid directly passes to the vapor phase.

2.2.4 Properties

Various properties are commonly used to define the state of a substance or flow of matter. They include:

- **Internal energy and specific internal energy.** Internal energy represents the molecular state type of energy. Specific internal energy is a measure per unit mass of the internal energy of a simple system in equilibrium. The specific internal energy of a liquid and vapor mixture is a function of the specific internal energies of the liquid and vapor, as well as the vapor quality.
- **Enthalpy and specific enthalpy.** Enthalpy represents another measure of energy, which typically includes internal energy and flow work. Specific enthalpy is a measure per unit mass of enthalpy. The specific enthalpy of a liquid and vapor mixture is a function of the specific enthalpies of the liquid and vapor, as well as the vapor quality.
- **Entropy and specific entropy.** Entropy is a measure of the molecular disorder of a substance at a given state. The specific entropy of a liquid and vapor mixture is a function of the specific entropies of the liquid and vapor, as well as the vapor quality.

2.2.5 Sensible and Latent Processes

The thermal capacity of a substance is a measure of its ability to hold or contain heat. When heat is added to or removed from a substance, two main types of processes can occur:

- **Sensible.** A sensible process is one in which the addition or removal of heat causes only temperature change. The heat absorbed or released by a substance during the change in temperature is referred to as sensible heat. An example is the heat absorbed by a liquid in raising its temperature to its boiling point.
- **Latent.** A latent process is one in which the addition or removal of heat causes only a phase change of the substance (e.g., boiling, condensation, melting, freezing). The heat absorbed or released in changing the phase of the substance is referred to as

latent heat. An example is the heat addition required to convert a liquid to a vapor at a fixed temperature and pressure.

A vapor is a gas that is at or near equilibrium with the liquid phase. In practice, this means a gas under or only slightly outside of the saturation curve. Vapor quality is the ratio of the mass of the vapor to the total mass of the liquid and vapor in a vapor mixture. A superheated vapor is a saturated vapor which has been heated so that its temperature is above the boiling point.

Most pure substances exhibit a specific melting and freezing temperature, and this is fairly independent of pressure. For example, water begins to freeze at 0 °C. The heat required to freeze water at 0 °C to ice at the same temperature, or vice versa, is called the latent heat of fusion of water, which is normally expressed on a per unit mass basis.

2.2.6 Ideal and Real Gases

In many practical situations, gases (e.g., hydrogen) and gaseous mixtures (e.g., air) can be treated as ideal gases. This is especially so for temperatures much higher than their critical temperatures, and for pressures much lower than their saturation pressures at a given temperature. Most gases and vapors exhibit ideal gas behavior at low pressures. For instance, low-pressure water vapor behaves as an ideal gas, and water vapor in the atmosphere, which exerts a low pressure (far below 1 atm), also exhibits ideal gas behavior. The ideal gas equation of state was developed empirically based on experimental measurements.

Real gases are often approximated as ideal. An ideal gas can be described via a simple equation of state:

$$Pv = RT \tag{2.1}$$

where v denotes specific volume, P pressure, T temperature, and R the gas constant. The gas constant R varies with gas type, and can be expressed as a function of its molecular weight M and the universal gas constant $\overline{R}$, which has a value of 8.314 kJ/mol K:

$$R = \frac{\overline{R}}{M} \tag{2.2}$$

Equation (2.1) can be expressed on a mole basis as:

$$P\overline{v} = \overline{R}T \tag{2.3}$$

or

$$PV = n\overline{R}T \tag{2.4}$$

For ideal gases, the specific heats at constant pressure and at constant volume are constants. This allows changes in specific internal energy and specific enthalpy during a process to be expressed as follows:

$$\Delta u = (u_2 - u_1) = c_v(T_2 - T_1) \tag{2.5}$$
$$\Delta h = (h_2 - h_1) = c_p(T_2 - T_1) \tag{2.6}$$

By noting that $h = u + Pv = u + RT$, it can be shown for ideal gases that

$$c_v - c_p = R \tag{2.7}$$

For the range of states encountered in many processes, the ideal gas model may not be adequately accurate. For real gases, other equations of state may be applicable for accurately representing the $P-v-T$ behavior of a gas, including the Benedict–Webb–Rubin, the van der Waals, and the Redlich and Kwong equations of state. These equations of state are more complex than that for an ideal gas. Another approach which extends the ideal gas equation of state is to use the compressibility factor Z to account for the deviation of a real gas from the ideal gas equation of state. The compressibility factor can be written as follows:

$$Z = \frac{Pv}{RT} \tag{2.8}$$

The ideal gas equation of state [Equation (2.1)] applied when $Z = 1$.

In some cases, one of the parameters P, v, or T is constant. At a fixed temperature, the volume of a quantity of ideal gas varies inversely with the pressure exerted on it. Then, for a process from initial state 1 to final state 2:

$$P_1 V_1 = P_2 V_2 \tag{2.9}$$

If temperature increases with compression at constant pressure, the volume of an ideal gas varies directly with absolute temperature as follows:

$$\frac{V_1}{T_1} = \frac{V_2}{T_2} \tag{2.10}$$

However, if temperature increases at constant volume, the pressure of an ideal gas varies directly with absolute temperature as follows:

$$\frac{P_1}{T_1} = \frac{P_2}{T_2} \tag{2.11}$$

If both temperature and pressure change simultaneously during a process, a combined ideal gas equation can be written:

$$\frac{P_1 V_1}{T_1} = \frac{P_2 V_2}{T_2} \tag{2.12}$$

The entropy change of an ideal gas can be expressed using the above equations as:

$$s_2 - s_1 = c_{v0} \ln\left(\frac{T_2}{T_1}\right) + R \ln\left(\frac{v_2}{v_1}\right) \tag{2.13}$$

or

$$s_2 - s_1 = c_{p0} \ln\left(\frac{T_2}{T_1}\right) - R \ln\left(\frac{P_2}{P_1}\right) \tag{2.14}$$

For a reversible polytropic process, the ideal gas parameters can be related with a polytropic exponent n. Then, $Pv^n = $ constant and we can write for a process:

$$Pv^n = P_1 v_1^n = P_2 v_2^n \tag{2.15}$$

and

$$\frac{P_2}{P_1} = \left(\frac{T_2}{T_1}\right)^{n/n-1} = \left(\frac{v_1}{v_2}\right)^n = \left(\frac{V_1}{V_2}\right)^n \tag{2.16}$$

Depending on the value taken on by n, four types of polytropic processes for ideal gases can be observed: isobaric ($n = 0$), isothermal ($n = 1$), isentropic ($n = k$), and isochoric ($n = \infty$). Here, k denotes the adiabatic exponent, expressible as the specific heat ratio:

$$k = \frac{c_{p0}}{c_{v0}} \tag{2.17}$$

For a closed system containing an ideal gas undergoing a polytropic process with constant specific heats, the work produced can be expressed using the first law of thermodynamics (described subsequently) as follows:

$$W_{1-2} = \frac{mR(T_2 - T_1)}{1 - n} = \frac{(P_2 V_2 - P_1 V_1)}{1 - n} \tag{2.18}$$

2.2.7 Energy and Power

Energy can appear in various forms as it is transferred:

- mechanical (i.e., work)
- electrical
- thermal (i.e., heat)
- chemical

Work is energy transfer due to a difference in pressure or the action of a force. Work can be divided into two types:

- **Shaft work.** This is a mechanical energy generated by a device (e.g., turbine) or used to drive a device (e.g., pump, compressor).
- **Flow work.** This is the energy transferred into or out of a system by the fluid flowing into or out of it.

Heat is a form of energy, specifically thermal energy. Heat transfer requires a temperature difference, and the higher the temperature difference, the greater the heat transfer rate. Thermal energy flows only from a higher temperature level to a lower temperature unless external energy is added to force the opposite to occur. If there is no heat transfer involved in a process, it is called adiabatic.

The rate of energy transfer, that is, the energy transfer per unit time, is referred to as power.

2.2.8 The Laws of Thermodynamics

There are four basic laws of thermodynamics: zeroth, first, second, and third. The first and second laws of thermodynamics are the most commonly employed in energy assessments, and are discussed here.

The first law of thermodynamics expresses the principle of conservation of energy, stating that energy can be neither be created nor destroyed in a closed system. This law accounts for all forms of energy (e.g., kinetic, potential, internal).

The second law of thermodynamics expresses the principle of non-conservation of entropy. Two classical statements of the second law follow:

- **Clausius statement.** It is impossible to construct a device, operating in a cycle without any external force, which transfers heat from a low-temperature to a high-temperature body.

- **Kelvin–Plank statement.** It is impossible to construct a device, operating in a cycle, which accomplishes only the extraction of heat from some source and its complete conversion to work.

Entropy is a measure of degree of disorder. Relatedly, all real energy transfers or conversions are irreversible. The second law of thermodynamics also states that the entropy of the universe is always increasing, that is, the sum of the entropy changes of a system and that of its surroundings must always be positive. The second law of thermodynamics helps determine the direction of processes, establish conditions for equilibrium, and determine quantitatively the best performance of a thermodynamic system as well as factors that preclude achieving the best theoretical performance.

2.2.9 Reversibility and Irreversibility

Reversibility and irreversibility are important concepts that relate to the performance, efficiencies and losses of thermodynamic processes and systems. Processes can be either reversible or irreversible:

- **Reversible process.** A reversible process is one in which both the system and its surroundings can be returned to their initial states. A reversible process is only a theoretical construct in that it presumes no losses of any kind.
- **Irreversible process.** An irreversible process is one in which both the system and its surroundings cannot be returned to their initial states, without the input of additional energy. The system incurs losses due to irreversibilities such as heat transfer across finite differences, friction, electrical and mechanical resistances, and unconstrained expansion.

Irreversibilities lead to a creation of entropy (or a destruction of exergy, as described in the next section). Thus, in reversible processes, entropy and exergy are conserved, while in irreversible processes, these quantities are not conserved. Rather, overall entropy increases and overall exergy decreases in irreversible processes. All real processes are irreversible, because there always exist irreversibilities in the real world.

Let us consider some examples to illustrate simply the differences in reversible and irreversible processes:

- **Electrical generating plant** (Figure 2.3). In an actual electrical generating plant, the actual electricity generated is always less than the reversible electricity generation rate, which is higher and which represents an ideal or maximum. The difference between the ideal and actual electricity generation rates is directly related to the irreversibility of the process.
- **Electrical heat pump** (Figure 2.4). Similarly, for an electrical heat pump, the actual electricity required to operate the device is always greater than the reversible electricity input rate, which is lower and represents an ideal or minimum. Again, the difference between the actual and ideal electricity input rates is directly related to the irreversibility of the process.

2.2.10 Exergy and Exergy Analysis

In this section, we describe the quantity exergy and the methodology derived from it, exergy analysis. Exergy and exergy analysis have been investigated and applied

Figure 2.3 Schematic of an electrical generating plant.

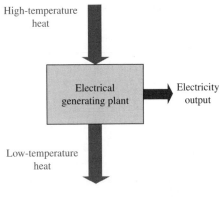

Figure 2.4 Schematic of an electrical heat pump.

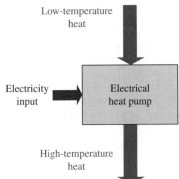

extensively (Dincer 1998; Dincer and Rosen 1999; Rosen 1999; Rosen and Dincer 2003, 2004; Cohce *et al.* 2011; Hacatoglu *et al.* 2011; Dincer and Rosen 2012; Kotas 2012; Aghbashlo *et al.* 2013; Park *et al.* 2014; Khalid *et al.* 2015).

2.2.10.1 Exergy

Exergy, also sometimes called availability, available energy, or essergy, is a thermodynamic quantity that is defined as the maximum amount of work producible from a stream of matter or energy (e.g., electricity, work, heat) as it comes to equilibrium with a reference environment.

When evaluating exergy, the state of the reference environment, or the reference state, must be specified. This is commonly done by specifying the temperature, pressure, and chemical composition of the reference environment.

Less rigorously, exergy is considered a measure of the potential to cause change of a stream of matter or energy, as a consequence of it not being in complete equilibrium the reference environment (i.e., not being completely stable relative to the reference environment). Exergy is also viewed as a measure of usefulness, quality, or value.

Exergy is not subject to a conservation law, except for reversible processes which are only hypothetical. Rather, exergy is destroyed (or consumed) in any real process due to irreversibilities.

Exergy is sometimes confused with other quantities, such as availability (or available energy) and essergy. To avoid confusion, relations are described in Table 2.1 between exergy and these related quantities. Note that exergy is different from free energy. The

Table 2.1 Exergy, availability, and essergy.

Quantity	Expression	Comments
Exergy	$E + P_0 V - T_0 S - (E_0 + P_0 V_0 - T_0 S_0)$	Introduced by Darrieus (1930) and Keenan in 1932 as the availability in steady flow. Rant (1956) coined the term exergy as a special case of essergy (i.e., essence of energy)
Availability	$E + P_0 V - T_0 S - (E_0 + P_0 V_0 - T_0 S_0)$	Formulated by Keenan in 1941 as a special case of essergy (Keenan 1951)
Essergy	$E + P_0 V - T_0 S - \Sigma_i \mu_{i0} N_i$	Formulated for a special case by Gibbs in 1878, as reported in Gibbs (1961). Formulated generally in 1962. Changed from available energy to exergy in 1963, and from exergy to essergy by Evans in 1968

Source: Adapted from Szargut *et al.* (1988).

Helmholtz Free Energy ($E - TS$) and the Gibbs Free Energy ($E + PV - TS$) were introduced by von Helmholtz and Gibbs in 1873, as reported in Gibbs (1961), as Legendre transforms of energy as useful alternate criteria of equilibrium, and as special cases of measures of the potential work of systems.

2.2.10.2 Exergy Analysis

Exergy analysis is a thermodynamic analysis method based on exergy. The method uses the principles of conservation of mass and energy as well as the non-conservation principle for entropy. Thus, exergy analysis is based on both the first and second laws of thermodynamics.

Exergy analysis is used for the assessment, design and optimization of energy systems (e.g., electricity generation, heating, refrigeration, chemical processing, and fuel production) as well as other types of systems (e.g., refining, production of industrial gases).

Exergy analysis is particularly useful for determining meaningful efficiencies and thermodynamic losses. Specifically, exergy analysis provides efficiencies that are true measures of how nearly efficiency approaches the ideal or upper limit, and pinpoints the locations, types, and magnitudes of losses (inefficiencies). Thus, exergy analysis enhances understanding of these quantities and greatly assists efforts to improve the efficiency with which we use energy resources. Additionally, exergy methods can help determine if more efficient energy systems can be designed by reducing losses in existing systems, and the actual margin for improvement.

2.2.10.3 Exergy vs Energy

Exergy and energy are different, although they are similar in some ways and have the same units. For clarity, exergy and energy are compared in Table 2.2, focusing on thermodynamic aspects.

A particularly significant feature of exergy, relative to energy, is that exergy is a measure of quality and quantity, while energy is only a measure of quantity. The use of a high

Table 2.2 Exergy and energy.

Feature	Quantity	
	Exergy	Energy
Relation to laws of thermodynamics	Based on a combination of the first and second laws of thermodynamics (whether process is reversible or irreversible)	Based on the first law of thermodynamics.
Efficiency	Always provides a measure of approach to ideality	Does not generally provide a measure of approach to ideality
Losses	Losses are of two types: waste emissions that contain exergy (external exergy losses) and exergy destructions due to irreversibilities (internal exergy losses)	Losses include only waste emissions that contain energy
Reference environment dependence	Dependent on both parameters/properties of material or energy flow and reference environment properties	Dependent on parameters/properties of material or energy flow only. Independent of reference environment properties
Value at equilibrium with reference environment	Zero when in equilibrium with the reference environment	Not zero in general when in equilibrium with the reference environment

quality energy form to satisfy a low quality energy need is usually inefficient thermodynamically (even if the energy efficiency is high).

As an example, consider electrical- and fuel-based heating. The use of a fossil fuel or electricity for heating at moderate or low temperatures (e.g., in applications such as space heating or domestic hot water heating) is typically quite inefficient on an exergy basis. This is because a high quality resource is used to satisfy a low quality demand. For fossil fuel or electrical heating at low temperatures, the energy efficiency is typically 60–95% but the corresponding exergy efficiency is usually 5–10%, reflecting the mismatch between energy supply and demand qualities. Exergy analysis permits a better matching of energy sources and applications, avoiding the use of high quality energy for low quality tasks. Carrying the example further, exergy analysis suggests that, to improve the exergy efficiency, a lower quality energy source could be used for low-temperature heating, or a much more efficient device that requires much less electricity or fuel such as a heat pump could be used.

2.2.10.4 Exergy and the Environment

The observation that exergy is a measure of the potential of a stream to cause change, in a thermodynamic system, has been extended to the broader environment. That is, some consider exergy a measure of the potential of a stream to cause change in the environment. Although not rigorous, this relationship has led to much research on the potential of exergy to provide a loose measure of the potential of a stream (material or energy) to impact the environment.

2.3 Heat Transfer

Facets of heat transfer most relevant to geothermal energy systems and their applications are introduced in this section to provide the reader with a good grounding in heat transfer, which is important to understanding geothermal energy.

2.3.1 Exchange of Heat

Heat transfer is the conveyance of thermal energy from one point to another. Heat transfer extends from the second law of thermodynamics which states that, of itself, heat tends to flow from higher to lower temperature. In order to transfer heat, therefore, a driving force is required in the form of a temperature difference between the heat origin and heat end point. For example, heat will flow from the hotter end towards the colder end of a metal bar that is heated on one end.

To understand the physical phenomena and practical aspects of heat transfer, a knowledge is required of the basic laws, governing equations, and boundary conditions. Numerous generalized relationships used in heat transfer have been developed, and many of these are discussed in this section.

Processes involving heat transfer are commonplace. Examples include heating and cooling of liquids, gases or solids, evaporation of liquids, and the extraction of the energy release by exothermic chemical reaction. Many applications exist in industry (e.g., space heating and cooling, domestic hot water heating, process heating, food processing, refrigeration) (Dincer 1997).

2.3.2 Modes of Heat Transfer

There exist three main modes of heat transfer: conduction; convection; and radiation (Figure 2.5). Note that there are also mixed modes of heat transfer, and heat transfer can even occur by all three modes simultaneously. The main features of the three heat transfer modes are:

- **Conduction.** When a temperature gradient exists in a stationary medium, solid or fluid, heat transfer through the medium is by conduction.
- **Convection.** Convection involves heat transfer between a surface and a moving fluid at different temperatures.

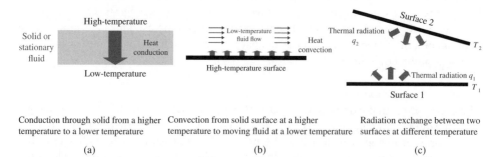

Conduction through solid from a higher temperature to a lower temperature	Convection from solid surface at a higher temperature to moving fluid at a lower temperature	Radiation exchange between two surfaces at different temperature
(a)	(b)	(c)

Figure 2.5 Illustrations of three modes of heat transfer: conduction (a); convection (b); and radiation (c).

Table 2.3 Common dimensionless heat transfer parameters.

Parameter	Expression
Biot number (Bi)	hY/k
Fourier number (Fo)	$\alpha t/Y^2$
Graetz number (Gz)	$(D/Y)\,\text{Re}\,\text{Pr}$
Grashof number (Gr)	$g\beta\Delta TY^3/\nu^2$
Rayleigh number (Ra)	$\text{Gr}\,\text{Pr}$
Nusselt number (Nu)	hY/k_f
Peclet number (Pe)	$UY/\alpha = \text{Re}\,\text{Pr}$
Prandtl number (Pr)	$c_p\mu/k = \nu/\alpha$
Reynolds number (Re)	UY/ν
Stanton number (St)	$h/\rho Uc_p = \text{Nu}/\text{Re}\,\text{Pr}$

- **Radiation.** Radiation is heat transfer occurring between two surfaces at different temperatures, with or without an intervening medium such as air, where the surfaces emit and absorb energy in the form of electromagnetic waves related to the temperatures of the surfaces.

Some standard dimensionless groups are commonly encountered in heat transfer (Table 2.3). One usually encounters the Biot number in situations involving conduction (steady-state or transient), and the Fourier number in situations involving unsteady-state conduction. The Grashof and Rayleigh numbers usually are encountered when natural convection occurs, while the Graetz number is common for laminar convection. Situations involving natural or forced convection, and boiling or condensation, often involve the Nusselt and Prandtl numbers. Forced convection situations often utilize the Reynolds, Stanton and Peclet numbers.

2.3.3 Conduction

Conduction is the transfer of heat from a substance to another part of the same substance or to another substance in physical contact, without significant displacement of the molecules forming the substances. In solids, conduction is due in part to the impact of adjacent molecules vibrating about their positions and internal radiation. When the solid is a metal, a large number of mobile electrons, which can easily move through matter, pass from one atom to another and help redistribute energy in the metal (an effect that is significant).

Conduction is governed by Fourier's law, which states that the instantaneous rate of heat flow through an individual homogeneous solid object is directly proportional to the cross-sectional area A (i.e., perpendicular to the direction of heat flow) and to the temperature difference driving force across the object with respect to the length of the path of the heat flow, dT/dx. For a thin solid slab of thickness dx and surface area A, with one face at a temperature T and the other at a lower temperature $(T - dT)$, Fourier's law implies the heat transfer rate can be expressed as:

$$Q = -kA\frac{dT}{dx} \tag{2.19}$$

where k denotes the thermal conductivity of the slab. The thermal conductivity is roughly constant for most solids over a broad range of temperatures. Integrating this equation from T_1 to T_2 for dT and 0 to L for dx yields:

$$Q = -k\frac{A}{L}(T_2 - T_1) = k\frac{A}{L}(T_1 - T_2) \tag{2.20}$$

2.3.4 Convection

Convection is fundamentally the transfer of heat by the motion or circulation of the heated parts of a liquid or gas. Convection occurs within a fluid by mixing different portions of it. Convection can be classified according to the nature of the flow:

- **Forced convection.** Forced convection occurs when a flow or motion of fluid is caused by mechanical or external means (e.g., fan, pump, wind) (Figure 2.6).
- **Natural or free convection.** Natural convection occurs when the flow is induced by buoyancy forces in the fluid caused by density variations, which are attributable to temperature variations in the fluid (Figure 2.7).

Heat transfer from a solid surface to a liquid or gas is partly by conduction and convection. Whenever there is appreciable movement of the fluid, heat transfer by convection dominates, although there normally is a thin boundary layer of fluid on a surface through which heat is transferred by conduction.

Convection is governed by Newton's law of cooling, which states that heat transfer from a solid surface to a fluid is proportional to the difference between the surface and fluid temperatures and the surface area. Thus, the convective heat transfer rate can be expressed as:

$$Q = hA(T_s - T_f) \tag{2.21}$$

where h denotes the convective heat transfer coefficient (or skin coefficient, or film coefficient). It accounts for all factors that affect convection, and depends mainly on the boundary layer conditions, surface geometry, fluid motion, and thermophysical properties. Note that radiation is not accounted for in this expression. In many problems, radiative heat transfer is much smaller than conductive and convective heat transfer, although this is not true in some instances, for example, radiation and convection contributions to heat transfer may both be significant when natural convection is the main form of convective heat transfer and/or when surface temperatures are high. In such

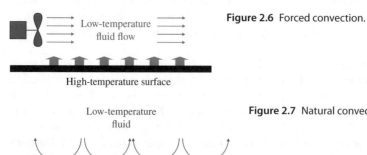

Low-temperature fluid flow

High-temperature surface

Figure 2.6 Forced convection.

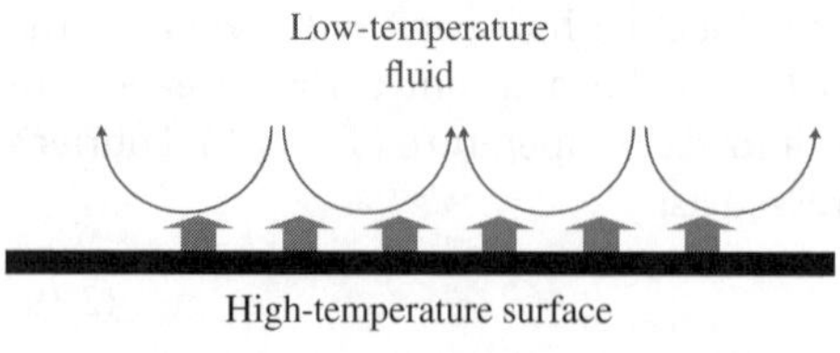

Low-temperature fluid

High-temperature surface

Figure 2.7 Natural convection.

instances, Equation (2.21) is modified to account for radiation contributions via the convective heat transfer coefficient, or a full radiative heat transfer assessment is carried out.

An important factor affecting heat transfer by convection is the nature of the flow, especially in the boundary layer. Convective heat transfer coefficients vary greatly for laminar and turbulent flows.

For better understanding, some details on forced convection are instructive. Forced convection provides much higher rates of heat transfer than natural convection, and is highly sensitive to heat transfer coefficient values. For a laminar boundary layer, fluid motion is highly ordered and smooth, while fluid motion is irregular and fluctuating in a turbulent boundary layer. Fluctuations enhance the heat transfer, but also increase surface friction. In a laminar boundary layer, transport is dominated by diffusion while in the turbulent region, transport is dominated by mixing. The Reynolds number can be used to determine the nature of a flow (laminar or turbulent or transition). Most heat exchangers utilize forced convection. It is difficult to obtain exact solutions to most forced convection problems, especially when turbulent flow is present, so various approximate solutions have been developed. Also, many types of heat transfer equipment are designed for forced convection, including forced air and forced water heaters, and forced air and forced water coolers.

Additional details on natural convection are also informative. Natural convection occurs due to differences in density and the action of gravity, which establish a natural circulation pattern that promotes heat transfer. For many problems involving fluid flow, natural convection is small, especially if forced convection is present. Heat transfer coefficients are generally significantly smaller for natural convection than forced convection. Natural convection is affected by viscous drag and thermal diffusion, so properties such as fluid kinematic viscosity and thermal diffusivity are important in such problems. These parameters depend on the fluid properties, the temperature difference between the surface and the fluid, and the characteristic length of the surface. Natural convection is often less expensive to facilitate than forced convection, which requires devices such as fans. Note that one must consider the radiative contribution to the total heat transfer in some cases, for example, natural convection and radiation heat transfer may be similar in magnitude for walls in buildings at room temperature.

2.3.5 Radiation

An object at any finite absolute temperature emits radiant energy in all directions and, if this radiation strikes another object, it may be transmitted and/or absorbed and/or reflected. This radiation heat transfer can transmit energy from a hot to a cold object. This is because, if two objects at different temperatures are located such that the radiation from each is intercepted by the other, the cooler object receives more energy than it radiates and vice versa, raising the internal energy of the cooler object and lowering it for the hotter one. The amount of energy radiated increases significantly (and non-linearly) as the temperature of a body rises. Radiation heat transfer occurs whether or not there is a medium (e.g., air) between the objects.

When radiation is incident on an object, the fraction absorbed by the object is called the absorptivity a, the fraction reflected is called the reflectivity r, and the fraction transmitted is called the transmissivity t. To account for all radiation incident on an object:

$$a + r + t = 1 \tag{2.22}$$

The transmitted radiation is negligibly small for many practical applications involving solids and liquids so Equation (2.22) is often simplified by setting the transmissivity to $t = 0$.

Although radiative heat transfer between solid surfaces is common, gases also emit and receive radiation. But while solids emit radiation over a wide range of wavelengths, some gases emit and absorb radiation at specific wavelengths.

Real materials (solids, liquids, and gases) have radiative properties that depend on the nature of the material.

For a blackbody, which is a special type of body whose properties are such that it absorbs all radiation, we can write the absorptivity as $a = 1$, Then, it is clear from Equation (2.22) that the reflectivity is $r = 0$ and the transmissivity is $t = 0$ for a blackbody.

The non-linearity with temperature of radiation is an outcome of the Stefan–Boltzmann law, which states that the emissive power of a blackbody E_b is directly proportional to the fourth power of its absolute temperature. That is,

$$E_b = \sigma T_s^4 \tag{2.23}$$

where T_s denotes the absolute temperature of the body and σ is the Stefan–Boltzmann constant (5.669×10^{-8} W/m² K⁴).

The Stefan–Boltzmann law enables the amount of radiation emitted in all directions and over all wavelengths to be determined based on the temperature of the blackbody. For a non-blackbody, the energy emitted E_{nb} can be expressed by extending Equation (2.23) as follows:

$$E_{nb} = \varepsilon \sigma T_s^4 \tag{2.24}$$

where ε denotes the emissivity of the body. With Equation (2.24), the radiative heat transferred per unit area from the surface of a body to its surroundings can be written as:

$$q = \varepsilon \sigma (T_s^4 - T_a^4) \tag{2.25}$$

Equation (2.25) is not valid if the emissivity ε of the body at absolute temperature T_s differs from that for the body (or surroundings) at T_a. In this case, the approximation is often applied in which the absorptivity of an object when receiving radiation from a source at T_a is taken to be equal the emissivity of the object when emitting radiation at T_a. With this approximation, we can write:

$$q = \varepsilon_{Ts} \sigma T_s^4 - \varepsilon_{Ta} \sigma T_a^4 \tag{2.26}$$

It is sometime convenient to express the net radiation heat transfer in the following form:

$$Q = h_r A (T_s - T_a) \tag{2.27}$$

where h_r denotes the radiation heat transfer coefficient. This form allows radiation heat transfer to be combined with convective heat transfer, since both equations have the same form. But the radiation heat transfer coefficient exhibits a significant temperature dependence while the convection heat transfer coefficient only exhibits a weak dependence. This can be seen with Equation (2.26) and Equation (2.27), which can be used to show that the radiation heat transfer coefficient has the following complex functional form:

$$h_r = \varepsilon \sigma (T_s + T_a)(T_s^2 + T_a^2) \tag{2.28}$$

2.3.6   Heat Transfer for Selected Simple Geometries

Heat transfer can be determined for most practical situations, although these can involve complex geometries. Analytical expressions for steady-state heat transfer can be developed for simple geometries, and some are considered here. No heat generation is presumed within the geometries.

Consider first a simple wall of thickness L. Heat transfer occurs through the wall from a high-temperature fluid A on one side of the wall to a low-temperature fluid B on the other side. In fluid A, the temperature decreases rapidly from T_A to the side 1 surface temperature T_{s1} near the wall in a thin boundary layer or film. A similar phenomenon occurs for fluid B on side 2. The steady-state, one-dimensional heat transfer per unit surface area q from fluid A to fluid B can be expressed as follows:

$$q = \frac{(T_A - T_B)}{(1/h_A + L/k + 1/h_B)} \tag{2.29}$$

Here, h denotes the convective heat transfer coefficient and k thermal conductivity. Note that $[(1/h_A) + (L/k) + (1/h_B)]^{-1}$ is referred to as the overall heat transfer coefficient, which itself is a function of other heat transfer coefficients.

Next, consider a composite wall having many layers of different materials. Each wall layer provides a separate thermal resistance. The one-dimensional heat transfer rate for such a composite wall or area A can be written as:

$$Q = \frac{(T_A - T_B)}{\Sigma R_t} = \frac{\Delta T}{\Sigma R_t} \tag{2.30}$$

where $\Sigma R_t = R_{t,t} = 1/HA$ and denotes the thermal resistance, while H denotes the overall heat transfer coefficient, which can be expressed as follows:

$$H = \frac{1}{R_{t,t}A} = \frac{1}{(1/h_1 + L_1/k_1 + \ldots + 1/h_n)} \tag{2.31}$$

Another practical but simple geometry is the hollow cylinder (e.g., a pipe). For a cylinder of internal radius r_1 and external radius r_2, whose inner and outer surfaces are subjected to fluid flows at different temperatures, the governing heat conduction equation can be written in a steady-state form as:

$$\frac{1}{r}\frac{d}{dr}\left(kr\frac{dT}{dr}\right) = 0 \tag{2.32}$$

By solving Equation 2.32 under appropriate boundary conditions and assuming k is constant, the following heat transfer equation is obtained:

$$Q = \frac{k(2\pi L)(T_1 - T_2)}{\ln(r_1/r_2)} = \frac{(T_1 - T_2)}{R_t} \tag{2.33}$$

2.4   Fluid Mechanics

A good grounding of fluid mechanics helps in understanding geothermal energy systems, as fluid flow problems often arise. Facets of fluid mechanics relevant to geothermal energy systems and their applications are introduced in this section to assist the reader.

2.4.1 Fluid Flow

Fluid flow is the movement of a fluid. The movement can be in an open space, within a conduit or over a surface (Figure 2.8).

As discussed in the previous section, a fluid flow may be laminar (i.e., orderly and smooth) or turbulent (i.e., fluctuating and chaotic) (Figure 2.9). This typically depends on disturbances, irregularities, or fluctuations in the flow, and is closely related to a fluid's velocity and viscosity. Laminar flow is common for fluids having low velocities and high viscosities while turbulent flow is common for fluids having high velocities and low viscosities. Between these two flow types, there is often a transition region where characteristics of both flow types are evident. The Reynolds number (Re), which indicates the ratio of inertia force to viscous force, can be expressed as:

$$\text{Re} = \frac{VD}{\upsilon} = \frac{\rho VD}{\mu} \tag{2.34}$$

and is commonly used to distinguish between laminar and turbulent flows. For example, fluid flow in a circular pipe is laminar when Re<2100, turbulent when Re>4000 and in the transition region for intermediate values of Reynolds number.

Viscosity represents a fluid's resistance to deformation. In gases, the viscosity increases with increasing temperature, resulting in greater molecular activity and

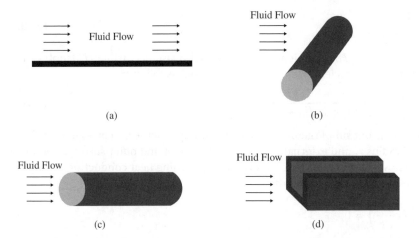

Figure 2.8 Flow over various geometries. (a) External flow over a flat plate; (b) external crossflow over a cylinder; (c) internal flow through a pipe; (d) internal flow through a channel.

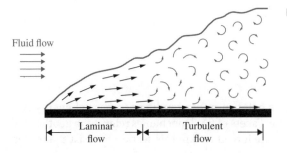

Figure 2.9 Laminar and turbulent flows.

momentum transfer. Molecular cohesion considerably affects the viscosity in fluids. Viscosity decreases with increasing temperature since the cohesive forces decrease. All real fluids are viscous, but an ideal fluid is inviscid, that is, it exhibits no shear stresses. There are two types of viscosities:

- **Dynamic viscosity.** The dynamic viscosity μ is the ratio of a shear stress to a fluid strain (velocity gradient).
- **Kinematic viscosity.** The kinematic viscosity v is defined as the ratio of dynamic viscosity to density ρ.

Two types of fluids can be classified based on viscosity:

- **Newtonian.** For a Newtonian fluid, the dynamic viscosity is dependent on temperature and pressure but independent of the velocity gradient magnitude. Examples include water and air.
- **Non-Newtonian.** Non-Newtonian fluids do not exhibit Newtonian behavior, but rather display more complex viscous characteristics. Examples include slurries, polymer solutions, and sludge.

Fluids flow in three dimensions in general, but flows are sometimes simplified to two- or one-dimensional:

- **Two-dimensional flow.** Fluid flow can be treated as two-dimensional if all fluid and flow parameters (e.g., velocity, pressure, temperature, density, viscosity) vary spatially in two dimensions.
- **One-dimensional flow.** Fluid flow can be treated as one-dimensional if all fluid and flow parameters vary only in one dimension (usually in the direction of flow).

In general, a fluid can be compressed. Two important fluid flow descriptions arise from this characteristic:

- **Compressible flow.** Fluid density changes with pressure for fluids that are compressible.
- **Incompressible flow.** Incompressible flow exists in situations where changes in fluid density with pressure are extremely small, and thus negligible.

Fluid flow may be uniform or non-uniform. For uniform flow, the velocity and cross-sectional area are constant in the direction of flow. If these quantities are not fixed, the flow is non-uniform.

The characteristics of a fluid flow may or may not be time varying. This leads to two distinct and important classifications of fluid flow:

- **Steady flow.** In steady flow, the flow conditions do not change with time.
- **Unsteady flow.** In unsteady flow, the fluid conditions vary with time.

For a steady flow in which conditions such as flow velocity, pressure and cross section vary spatially but not temporally, the flow is steady but non-uniform. This contrasts with steady uniform flow, where all conditions are uniform and unvarying with time or position. We can treat unsteady flows similarly. In an unsteady uniform flow, the velocity at every location in the flow field is the same but changing with time, for example, an accelerating or decelerating flow. But where the cross-sectional area and velocity vary temporally and from point to point, unsteady uniform flow exists.

Liquids flowing in open channels may also be classified as subcritical, critical, or super-critical depending on their flow regions.

Gaseous flows may be categorized by speed as subsonic, transonic, supersonic, or hypersonic.

2.4.2 Governing Equations

The basic governing equations for fluid flow derive from such general fundamental principles as conservation of mass, conservation of motion (i.e., Newton's second law of motion), and conservation of energy.

2.4.2.1 Continuity Equation

The continuity equation is based on the principle of conservation of mass, and relates the velocity u and density ρ in a flowing fluid. For steady flow, the rate of mass change is zero, so the mass of fluid in a control volume remains constant and the masses of fluid entering and exiting per unit time are equal. For steady flow in a stream tube, the continuity equation for a compressible fluid is:

$$\rho_1 A_1 u_1 = \rho_2 A_2 u_2 = \text{constant} \tag{2.35}$$

where u_1 and u_2 are the velocities at Sections 1 and 2, respectively. Here, $\rho_1 A_1 u_1$ and $\rho_2 A_2 u_2$ are the mass rates entering at Point 1 and exiting at Point 2, respectively.

For incompressible fluids $\rho_1 = \rho_2$, and Equation (2.35) reduces as follows:

$$A_1 u_1 = A_2 u_2 = \text{constant} \tag{2.36}$$

Here, $A_1 u_1$ and $A_2 u_2$ are the volumetric flow rates entering at Point 1 and exiting at Point 2, respectively. The unsteady-state continuity equation for an incompressible fluid in a stream tube is:

$$\left(\frac{dm}{dt} \right)_{sys} = \frac{d}{dt} \int_{cv} \rho dV + \int_{cs} \rho \overline{V} d\overline{A} \tag{2.37}$$

In Cartesian coordinates, the steady- and unsteady-state continuity equations, respectively, for an incompressible fluid can be written as follows:

$$\frac{\partial u}{\partial x} + \frac{\partial v}{\partial y} + \frac{\partial w}{\partial z} = 0 \tag{2.38}$$

$$\frac{\partial(\rho u)}{\partial x} + \frac{\partial(\rho v)}{\partial y} + \frac{\partial(\rho w)}{\partial z} = \frac{\partial \rho}{\partial t} \tag{2.39}$$

The above equation reduces to the steady-state continuity equation for a compressible fluid by setting the right side to zero. In cylindrical coordinates, which are common in piping systems, the steady-state continuity equations for incompressible and compressible fluids, respectively, are:

$$\frac{\partial v_r}{\partial r} + \frac{1}{r}\frac{\partial v_\theta}{\partial \theta} + \frac{\partial v_z}{\partial z} + \frac{v_r}{r} = 0 \tag{2.40}$$

$$\frac{\partial(\rho v_r)}{\partial r} + \frac{1}{r}\frac{\partial(\rho v_\theta)}{\partial \theta} + \frac{\partial(\rho v_z)}{\partial z} + \frac{\rho v_r}{r} = 0 \tag{2.41}$$

The non-steady-state continuity equations for a compressible fluid can be written in Cartesian and cylindrical coordinates, respectively, as follows:

$$\frac{\partial \rho}{\partial t} + \frac{\partial(\rho u)}{\partial x} + \frac{\partial(\rho v)}{\partial y} + \frac{\partial(\rho w)}{\partial z} = 0 \tag{2.42}$$

$$\frac{\partial \rho}{\partial t} + \frac{\partial(\rho v_r)}{\partial r} + \frac{1}{r}\frac{\partial(\rho v_\theta)}{\partial \theta} + \frac{\partial(\rho v_z)}{\partial z} + \frac{\rho v_r}{r} = 0 \tag{2.43}$$

2.4.2.2 Momentum and Euler Equations

The analysis of fluid flow is governed in part by Newton's second law of motion, which generally states that when the net external force acting on a system is zero, the linear momentum of the system in the direction of the force is conserved in both magnitude and direction. This conservation of linear momentum is applicable to all types of fluid flows (e.g., steady, unsteady, compressible, incompressible). The motion of a fluid particle is described relative to an inertial coordinate frame.

The one-dimensional momentum equation at constant velocity is expressible as follows:

$$\Sigma F = \frac{d}{dt}(mV) \tag{2.44}$$

Here, ΣF represents the sum of the external forces acting on the fluid, and mV represents the kinetic momentum in that direction. Hence, Equation (2.44) expresses the equality between the time rate of change of the linear momentum of the system in the direction of V and the resultant of all forces acting on the system in the direction of V. The complete linear momentum equation is a vector equation and is dependent on a set of coordinate directions. For a steady flow with a constant velocity across the control surface, the momentum equation in scalar form becomes:

$$\Sigma F_x = (\dot{m}V_x)_e - (\dot{m}V_x)_i \tag{2.45}$$

or, if $\dot{m}$ is constant,

$$\Sigma F_x = \dot{m}(V_{xe} - V_{xi}) \tag{2.46}$$

Note that the rate of change of momentum of a control mass system can be related to the rate of change of momentum of a control volume using the continuity equation. Similar expressions as Equation (2.45) and Equation (2.46) be written for directions y and z.

The Euler equation is a statement of Newton's second law of motion, and can be applied to an inviscid fluid continuum. The equation states that the product of mass and acceleration of a fluid particle can be equated in a vector sense with the external forces acting on the particle.

The following simplified form of Euler's equation can be written for a steady flow along a stream tube:

$$\frac{1}{\rho}\frac{dp}{ds} + v\frac{dv}{ds} + g\frac{dz}{ds} = 0 \tag{2.47}$$

where p denotes pressure, v velocity, ρ density and z elevation. For an incompressible fluid, ρ is constant and the above equation can be integrated along the streamline s for an inviscid fluid to show that:

$$\frac{p}{\rho} + \frac{v^2}{2} + gz = \text{constant} \tag{2.48}$$

For a compressible fluid, Equation (2.47) can only be integrated as follows:

$$\int \frac{dp}{\rho g} + \frac{v^2}{2g} + z = H \tag{2.49}$$

The relationship between ρ and p is required. This relationship is often of the form $\rho(dp/d\rho) = K$ for liquids, where K is an adiabatic modulus, and $p\rho^n = $ constant for gases (which covers processes ranging from adiabatic to isothermal).

2.4.2.3 Bernoulli and Navier–Stokes Equations

The equations of motion may divided into two general types for inviscid (i.e., frictionless) fluids and for viscous fluids. These considerations lead to the Bernoulli and Navier–Stokes equations.

The Bernoulli equation can be written for both incompressible and compressible flows. The Bernoulli equation for incompressible flow is sometimes referred to as a mechanical-energy equation because of the similarity between it and the steady-flow energy equation derived from the first law of thermodynamics for an inviscid fluid with no heat transfer or external work. Note that for inviscid fluids the viscous forces and surface tension forces are not accounted for, so it is only valid when viscous effects are negligibly small. The general Bernoulli equation per unit mass for inviscid fluids between any two points is:

$$\frac{u_1^2}{2g} + \frac{p_1}{\rho g} + z_1 = \frac{u_2^2}{2g} + \frac{p_2}{\rho g} + z_2 = H \tag{2.50}$$

where $u^2/2g$ represents the kinetic energy per unit mass (or the velocity head), $p/\rho g$ denotes the pressure energy per unit mass (or the pressure head), z denotes the potential energy per unit mass (or the potential head or constant total head), H denotes the total energy per unit mass (or the total head), and subscripts 1 and 2 denote points along the streamline. The terms in Equation (2.50) represent energy per unit mass and have the unit of length. A comparison of Equation (2.50) with the general energy equation indicates that the Bernoulli equation contains restrictions linked to its main assumptions: the flow is steady, incompressible, and frictionless along a single streamline, and no shaft work and heat interactions occur.

The Bernoulli equation is used in many applications, particularly for flows with negligibly small losses, for example, in hydraulic systems.

The Navier–Stokes equations are differential expressions of Newton's second law of motion, and are constitutive equations for viscous fluids. For viscous fluids, two forces are considered: body force and pressure force on the surface. The solutions of these equations depend on what information is known and often must be obtained numerically. Nonetheless, exact solutions to the non-linear Navier–Stokes equations are known for some simple cases, for example, steady, uniform flows (either two-dimensional or with radial symmetry) and flows with simple geometries. Approximate solutions can be obtained for some one-dimensional simple flow cases that require only momentum and continuity equations in the flow direction, for example, uniform flow between parallel plates, uniform free surface flow down a plate, and uniform flow in a circular tube.

2.4.3 Pipe Flow

Pipe flow is a particularly common application in geothermal energy systems. Many empirical pipe-flow equations have been developed, particularly for water. The Hazen–Williams equations are commonly utilized. With these equations, the velocity V can be written as:

$$V = 0.850CR_h^{0.63}S^{0.54} \tag{2.51}$$

and the volumetric flow rate $\dot{V}$ as:

$$\dot{V} = 0.850CR_h^{0.63}S^{0.54}A \tag{2.52}$$

Here, A is the pipe cross-sectional area, R_h is the hydraulic radius of the pipe, S is the slope of the total head line, h_f/L, and C is the roughness coefficient, which varies with type of pipe. For badly corroded iron or steel pipes, for example, $C = 140$.

2.4.4 Boundary Layer

As a fluid moves past a solid like a wall, the relative velocity between the fluid and the solid surface is zero. This no-slip condition implies the velocity of the fluid particles at the solid equals the velocity of the solid, and is associated with viscous effects. Consequently, there exists a velocity gradient, which typically is non-linear, starting at zero at the solid and increasing into the flow until it reaches the bulk fluid flow velocity. The region in which a real fluid flow exhibits a reduced flow velocity is known as the boundary layer.

A boundary layer exists in both internal and external flows. In internal flow (e.g., pipe or open-channel flow), the boundary layer grows until the entire fluid is encompassed (Figure 2.10a). In external flows (e.g., flow along a plate or wall), the boundary layer continually grows because no confining boundary exists (Figure 2.10b). The free-stream velocity u_s is the velocity at a distance from the solid at which the fluid velocity is in essence unaffected by the presence of the solid.

The boundary layer thickness can be expressed in several ways. One simple approach is to determine the thickness at the point where the velocity u in the boundary layer approaches the free stream velocity u_s. For example, the boundary layer thickness δ is usually defined as the distance from the boundary to the point at which $u = 0.99u_s$.

For uniform flow of an incompressible fluid at a free-stream velocity towards a plate, the fluid has a zero velocity when it is in contact with the plate surface, that is, the no-slip condition holds. The boundary layer gets thicker in the direction of flow and the region with a velocity gradient grows with greater distance normal to the plate. The flow in the boundary layer typically starts out as laminar as the fluid encounters the plate, and then goes through a transition region before becoming turbulent some point downstream of where the fluid first encountered the plate. Thus, the rate of change of velocity determines both the velocity gradient at the surface and the shear stresses. Along the plate, the shear force gradually increases as the laminar boundary layer thickens, because of the increasing plate surface area affected. A turbulent boundary layer occurs after fluid instabilities materialize. The two boundary layer flow regimes (laminar and turbulent),

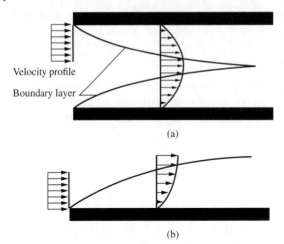

Velocity profile

Boundary layer

(a)

(b)

Figure 2.10 Boundary layer in internal and external flows. (a) Formation of velocity boundary layer for internal flow between two flat plates; (b) formation of velocity boundary layer for external flow over a flat plate.

and the transition from a laminar to a turbulent boundary layer, can be characterized by the Reynolds number. In pipe flow, for instance, the flow is laminar for Re<2300 and turbulent at higher Reynolds numbers. There is a limit to the growth of the boundary layer thickness in pipe flow because of the pipe size. For flow along a flat plate, a similar situation holds, except for different Reynolds number values and except for the fact that there is no limit to growth of the boundary layer.

2.5 The Nature of the Ground

In this section, basic concepts and background about the ground are presented since such material is fundamental to understanding ground-based geothermal systems and to provide a basis for material in later chapters.

2.5.1 Ground Composition

The earth has two major divisions. One division includes the metallic and silicate portions, and the other the core and the mantle (including the crust). The silicate portion of the earth is subdivided into the crust, the upper mantle, and the lower mantle, while the metallic earth is subdivided into an outer liquid shell and an inner solid region that is composed of Fe and Ni. The physical properties of the three main regions of the mantle are distinct. This, in principal, can be due to either compositional changes or phase changes with depth. All human activities related to geothermal energy within the earth are limited to the earth crust which extends down to about 30–50 km beneath continents and is mainly composed of rock.

The top layer of the earth's crust contains soil, which is a mixture of minerals, organic matter, gases, liquids, and organisms. Soil includes the loose rock material that lies above the solid geology of the earth. Soil provides a medium for plant growth and a habitat for organisms. It also carries out water storage, supply and purification. Soil continually undergoes change due to the many physical, chemical and biological processes to which it is subject, including weathering with erosion. Soil in fact results from the influence of the materials (original minerals) and organisms interacting over time in a climate. Soil formation is affected by the ground relief, which includes factors such as elevation, orientation, and slope. Soil particles can be classified by their chemical composition and size. The particle size distribution and mineralogy of a soil determines many of its properties. The mineralogy of fine soil particles, for example, clay, vary. The density varies between $1\,\mathrm{g/cm^3}$ and $2\,\mathrm{g/cm^3}$ for most soil types.

2.5.2 Groundwater

The rocks in the crust of the earth contain a network of pores and fractures that are often filled, to an extent, by groundwater. Such porous rocks, as well as other material such as sand and gravel, are called an aquifer when saturated with water. An aquifer can be used for groundwater extraction using water wells. Groundwater originates from precipitation on the ground surface, which percolates through the porous zone until it reaches the saturated zone. The top of this saturated zone is a surface called the water table, whose level fluctuates seasonally and from year to year depending on water transfers from precipitation and surface-water bodies. Therefore, the depth of this water table varies geographically in various climates, and is located from only a meter below the surface in humid climates to one hundred or more meters below the ground surface.

In the ground, groundwater moves from a higher elevation or pressure to a lower one. The movement of groundwater occurs at much lower speeds than surface water velocities and depends on the permeability of the body within which it is contained. While the flow of a surface stream is normally on the order of meters per second, groundwater can move as slowly as centimeters per year or even per decade (Alley *et al.* 1999). Groundwater can be naturally discharged to springs or along streams, lakes, and wetlands, or it can be discharged using water wells. Groundwater that reaches lower depths in the ground where hot molten rocks exist is heated and, if discharged, can be used in heating and electricity generation.

2.5.3 Ground Temperature Variations

Measurements show that, below a certain depth in the ground, the temperature fluctuations observed near the surface of the ground diminish (Figure 2.11), and the temperature remains relatively constant (e.g., at about 12 °C in various states in the USA) throughout the year (Geothermal Heat Pump Consortium 2013). This is due to the high thermal inertia of the earth surface and the time lag between the temperature fluctuations at the surface and their effect deeper in the ground. As a result, the heat from solar irradiation is not absorbed at very deep layers within the earth.

In addition to surface temperature variations, the temperature of the ground increases in direct proportion to increasing depth from the surface of the ground. This is called the geothermal gradient and is about 20–30 °C per kilometer of depth. The temperature difference between the earth's core and its surface drives the conduction of geothermal heat, which is produced by the radioactive decay of elements in deeper layers of the earth, towards the earth's surface. This heat can be used as an energy source for space heating and electricity generation. Active volcanoes and hot springs, distributed over various parts of the earth, are examples of heat that originates from layers deep within the earth.

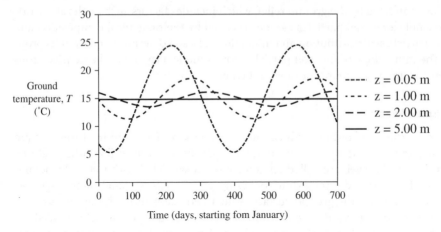

Figure 2.11 Ground temperature fluctuations with time for four values of depth *z*. *Source*: Adapted from Hillel (1982).

2.5.4 Soil Microbial Communities

Microorganisms are arguably the most diverse and abundant group of organisms on Earth. Soil contains a diverse range of bacteria that play an important role in soil formation, ecosystem biogeochemistry, contaminant degradation, and the maintenance of groundwater quality. The density of microorganisms are highest in the top layers of the soil surface and decrease with soil depth. Large numbers of microorganisms reside several meters from the soil surface.

Environmental changes in such quantities as soil temperature and soil moisture have been shown to influence soil microbial community composition. The effect of environmental changes or disturbances on the bacterial communities is an important topic in soil microbial ecology.

References

Aghbashlo, M., Mobli, H., Rafiee, S. and Madadlou, A. (2013) A review on exergy analysis of drying processes and systems. *Renewable and Sustainable Energy Reviews*, **22**, 1–22.

Alley, W.M., Reilly, T.E., and Franke, O.L. (1999) *Sustainability of Ground-Water Resources*, U.S. Geological Survey, Circ. 1186.

Cohce, M.K., Dincer, I. and Rosen, M.A. (2011) Energy and exergy analyses of a biomass-based hydrogen production system. *Bioresource Technology*, **102 (18)**, 8466–8474.

Darrieus, G. (1930) Définition du rendement thermodynamique des turbines a vapeur. *Revue Général de l'Electricité*, **27 (25)**, 963–968.

Dincer, I. (1997) *Heat Transfer in Food Cooling Applications*, Taylor & Francis, Washington, DC.

Dincer, I. (1998) Thermodynamics, exergy and environmental impact. *Proceedings of the ISTP-11, the Eleventh International Symposium on Transport Phenomena*, November 29–December 3, 1998, Hsinchu, Taiwan, pp. 121-125.

Dincer, I. and Rosen, M.A. (1999) Energy, environment and sustainable development. *Applied Energy*, **64** (**1–4**), 427–440.

Dincer, I. and Rosen, M.A. (2012) *Exergy: Energy, Environment and Sustainable Development*, 2nd edn, Elsevier, London.

Geothermal Heat Pump Consortium (2013) http://www.geoexchange.org (accessed January 29, 2016).

Gibbs, J.W. (1961) *The Scientific Papers of J. Willard Gibbs*, General Publishing Company, Toronto.

Hacatoglu, K., Dincer, I. and Rosen, M.A. (2011) Exergy analysis of a hybrid solar hydrogen system with activated carbon storage. *International Journal of Hydrogen Energy*, **36** (**5**), 3273–3282.

Hillel, D. (1982) *Introduction to Soil Physics*, Academic Press, San Diego.

Keenan, J. (1951) Availability and irreversibility in thermodynamics. *British Journal of Applied Physics*, **2**, 183–192.

Khalid, F., Dincer, I. and Rosen, M.A. (2015) Energy and exergy analyses of a solar-biomass integrated cycle for multigeneration. *Solar Energy*, **112**, 290–299.

Kotas, T.J. (2012) *The Exergy Method of Thermal Plant Analysis*, Paragon Publishing, New York.

Park, S.R., Pandey, A.K., Tyagi, V.V. and Tyagi, S.K. (2014) Energy and exergy analysis of typical renewable energy systems. *Renewable and Sustainable Energy Reviews*, **30**, 105–123.

Rant, Z. (1956) Exergy, a new word for technical available work. *Forschung im Ingenieurwesen*, **22**, 36-37 (in German).

Rosen, M.A. (1999) Second-law analysis of aquifer thermal energy storage systems. *Energy – The International Journal*, **24**, 167–182.

Rosen, M.A. and Dincer, I. (2003) Exergy–cost–energy–mass analysis of thermal systems and processes. *Energy Conversion and Management*, **44** (**10**), 1633–1651.

Rosen, M.A. and Dincer, I. (2004) Effect of varying dead-state properties on energy and exergy analyses of thermal systems. *International Journal of Thermal Sciences*, **43** (**2**), 121–133.

Szargut, J., Morris, D.R. and Steward, F.R. (1988) *Exergy Analysis of Thermal, Chemical, and Metallurgical Processes*, Hemisphere, New York.

3

Background and Technologies

3.1 Introduction

To analyze an energy system, an understanding is needed of the various technologies and system components that are often used in energy systems. This chapter provides background information on components relevant to geothermal energy systems such as heat pumps, heat exchangers, heating, ventilating, and air conditioning (HVAC) equipment, and energy storage units.

3.2 Heat Pumps

Heat pumps are cyclic devices that transfer heat from a low-temperature medium to a high-temperature medium. They consume electrical power to transfer heat against its natural flow direction. Heat pumps consist of a condenser, an expansion device, an evaporator, and a compressor (Figure 3.1). The working fluid is usually a type of refrigerant.

To examine the heat pump cycle, we start with the fluid that exits the condenser as a liquid at high pressure. The liquid refrigerant passes through an expansion device, which reduces the pressure of the refrigerant. The refrigerant at low pressure passes through a heat exchanger (evaporator) and absorbs heat from the low-temperature source. The refrigerant evaporates as heat is absorbed. The vapor refrigerant then passes through a compressor where it is pressurized, raising its temperature. The hot vapor then circulates through a condenser where heat is removed and passed to the high-temperature sink. As the refrigerant rejects heat, its phase changes back to liquid and the process begins again.

The objective of a heat pump that is in place for heating a space is to provide heat from the low-temperature environment to the heated space using electricity. The objective of a heat pump that is in place for cooling an environment is to remove heat from the cooled space to the high-temperature environment using electricity. Heat pumps can provide significant energy savings since they can transfer more energy than they consume. To evaluate the efficiency of a heat pump, or to compare various types of heat pumps, the coefficient of performance (COP_{HP}) is used, which is defined as follows:

Geothermal Energy: Sustainable Heating and Cooling Using the Ground, First Edition.
Marc A. Rosen and Seama Koohi-Fayegh.
© 2017 John Wiley & Sons, Ltd. Published 2017 by John Wiley & Sons, Ltd.

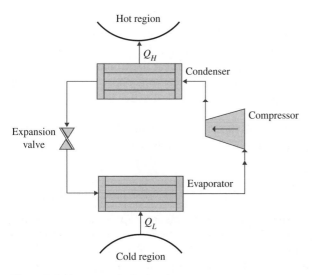

Figure 3.1 Heat pump in the heating season.

$$\text{COP}_{\text{HP,spaceheating}} = \frac{\text{Heat provided to heated space}}{\text{Electricity consumption}} \tag{3.1}$$

$$\text{COP}_{\text{HP,spacecooling}} = \frac{\text{Heat removed from cooled space}}{\text{Electricity consumption}} \tag{3.2}$$

The higher the COP, the more efficient the system. The temperatures of the high- and low-temperature media within which the heat pump operates directly affect the COP_{HP}. The COP_{HP} decreases at larger temperature differences between the high- and low-temperature media. A typical heat pump has a COP_{HP} of about four, that is, for every one unit of energy used to power the system, more than four units are delivered to the space as heat. Therefore, compared with efficiencies of 0.92 for a high efficiency natural gas furnace, heat pumps are considered very efficient for heating and cooling spaces.

The maximum efficiency that may be achieved by an "ideal" heat pump is defined by the theoretical "Carnot process" where all the processes are reversible. In an ideal heat pump, the coefficient of performance of the heat pump (COP_{rev}) is only dependent on the high- and low-temperature media. Specifically,

$$\text{COP}_{\text{rev}} = \frac{1}{1 - \frac{T_L}{T_H}} \tag{3.3}$$

The value of the COP deteriorates as the temperature difference between the heat sink and the heat source rises. Therefore, it is important to seek reasonable temperatures for the heat source and to reduce the temperature on the heat rejection side.

Heat pumps that are used for heating and cooling spaces have the space as their high-temperature and low-temperature medium, respectively. A variety of options such as air, water, and ground exist for the other medium from/to which heat is transferred. Waste heat sources from industrial processes, cooling equipment, or ventilation air from buildings can also be used for heating using heat pumps. Several factors affect the selection of such media, including their availability and temperature. Since the space is often kept at predefined temperatures, the temperature of the other medium greatly affects the COP_{HP} and capacity of the heat pump. When heating a space, the lower the temperature of the medium from which the heat is extracted, the higher is the rate of heat pump electricity consumption. When cooling a space, the higher the temperature of the medium to which the heat is delivered, the higher is the rate of heat pump electricity consumption.

Developments in heat pump design and in the compressors, motors, and controls that make up the heat pump are helping to improve heat pump efficiencies.

3.3 Heat Exchangers

Heat exchangers are devices that are used in many engineering applications, including space heating and cooling, for heat exchange processes between two media. The heat flows via a heat exchanger from a higher- to a lower-temperature medium. For example, the condenser in the heat pump cycle is a heat exchanger that transfers heat from the working fluid in the heat pump cycle to water that flows in the space heat distribution system. The rate of heat exchange depends directly on the temperature difference and heat transfer area between the media. Several heat exchanger configurations exist based on how the two media are brought into contact. The variation of the temperature of the two media over their heat transfer area, which depends on the heat capacity of the media, is often analyzed.

In designing heat exchangers, knowledge of parameters such as inlet and outlet temperatures of the flows, mass flow rates, and thermal properties of the flows at temperature ranges in which the heat exchanger operates helps in sizing the heat exchangers. Heat exchangers are designed as steady-flow and steady-state devices, although changes in flow rates and the thermal characteristics of the media frequently occur in real systems. When assessing the performance of heat exchangers or modeling their operation, the heat exchanger geometric characteristics and inlet flow conditions are often used to model their heat transfer and evaluate outlet temperatures. Where transient effects are important, computations related to the design and analysis of heat exchangers usually become complicated and numerical codes are used.

3.4 Heating, Ventilating, and Air Conditioning

HVAC systems provide heating, cooling, humidification, dehumidification, and ventilation for spaces. HVAC devices usually share common operating principles but often vary in physical appearance and arrangement and how they are controlled according to user needs. They often consist of a heating system, a cooling system, and a distribution system.

A schematic of the energy flow in space heating is shown in Figure 3.2. Heat is transferred from a heating system such as furnace or heat pump to the distribution system such as ducts via a heating coil. For cooling the space, heat is removed from the space and transferred to the cooling system such as chillers via a cooling coil. A schematic of the energy flow in space heating is shown in Figure 3.3. Heating, cooling and distribution systems are designed based on the space energy needs and the application.

Figure 3.2 Energy flows in space heating.

Figure 3.3 Energy flows in space cooling.

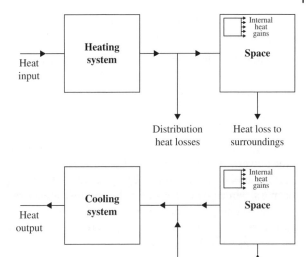

3.5 Energy Storage

Energy systems play a key role in harvesting energy from various sources and converting it to the energy forms required for applications in various sectors, for example, utility, industry, building, and transportation. Energy sources such as fossil fuels can be used to provide energy according to customer demand, that is, they are readily storable when not required. But other sources such as solar and wind energy need to be harvested when available and stored until needed.

Applying energy storage can provide several advantages for energy systems, such as permitting increased penetration of renewable energy and better economic performance. Also, energy storage is important to electrical systems, allowing for load leveling and peak shaving, frequency regulation, damping energy oscillations, and improving power quality and reliability.

Energy storage systems have been used for centuries and undergone continual improvements to reach their present levels of development. Numerous energy storage units are mature and commercial. Many types of energy storage systems exist, and they can be categorized in various ways such as their storage characteristics, storage period, and their applications. Some of the main types of energy storage are:

- battery storage
- hydrogen energy storage
- flywheel energy storage
- compressed-gas energy storage
- pumped storage
- magnetic storage
- chemical storage
- thermal energy storage
- thermochemical energy storage
- organic and biological energy storage

One categorization of these types of energy storage, based on the form in which the energy is stored, is as follows:

- chemical (including battery, hydrogen and thermochemical storage)
- thermal
- mechanical (including flywheel, compressed gas and pumped storage)
- magnetic
- biological and organic

Energy storage characteristics are often expressed in terms of specific energy and specific power which helps identify the potentials of each storage type and contrast them for various applications. Storage energy density is the energy accumulated per unit volume or mass, while power density is the energy transfer rate per unit volume or mass. When generated energy is not available for a long duration, a high energy density device that can store large amounts of energy is required. When the discharge period is short, as for devices with charge/discharge fluctuations over short periods, a high power density device is needed.

Energy storage systems can also be classified based on storage period. Short-term energy storage typically involves the storage of energy for hours to days, while long-term storage refers to storage of energy from a few months to a season (3–6 months). For instance, a long-term ground thermal energy storage retains thermal energy in the ground over the summer for use in winter.

Energy storage systems can be categorized according to various applications such as utilities, renewable energy utilization, buildings and communities and transportation. Hybrid energy storage (combining two or more energy storage types) is sometimes used, usually when no single energy storage technology can satisfy all application requirements effectively. Storage mass is often an important parameter in applications due to weight and cost limitations, while storage volume is important when the system is in a space-restricted or costly area such as an urban core. Energy storage applications are continuously expanding, often necessitating the design of versatile energy storage and energy source systems with a wide range of energy and power densities.

4

Underground Thermal Energy Storage

4.1   Introduction

This chapter provides background information on thermal energy storage (TES) concepts and theory, as well as thermal storage types, operation, and applications. Recent advances in thermal storage are noted and illustrative examples are presented of major thermal storage installations to demonstrate what has been achieved in the field.

Thermal storage entails the storage of thermal energy, whether heat (thermal energy above the ambient temperature) or cold (thermal energy below the ambient temperature). Both heat and cold storage are widely used.

The main thermal energy storage types are sensible, latent, and thermochemical. Energy is stored by raising the temperature of a storage medium in sensible heat storage, changing the phase of a storage medium in latent heat storage, and chemically reacting a storage medium in thermochemical storage.

Applications are numerous for some types of thermal storage (e.g., sensible) and rare for others (e.g., thermochemical). Nonetheless, TES has achieved noteworthy market penetration. All types of thermal storage have practical applications. The selection or design of storage for a given application is usually based – at least in part – on the comparative characteristics of applicable TES.

The design, operation, economics and testing of thermal storage vary from relatively straightforward for mature types of thermal storage to complex for other types. The design and operation become increasingly complex for intermittent sources of thermal energy (such as renewables like solar energy), as their temporal fluctuations must be matched with the independently time-varying demands for heating and cooling in applications such as building heating, ventilating, and air conditioning (HVAC) systems. TES can facilitate the management of such temporal mismatches between energy demand and supply.

This chapter goes on to describe TES methods, including types of thermal storage and their fundamentals, advantages, and operation. Additionally, factors affecting TES performance such as thermal energy quality and stratification are discussed. The economics and design of TES are also covered, as are relevant markets and applications.

Underground TES is described in greater detail in this chapter in a separate section, given its relevance to this book. The coverage includes information on types and characteristics of underground thermal storage. Finally, material is provided on the integration of heat pump technology with TES, as that often forms the basis of geothermal

Geothermal Energy: Sustainable Heating and Cooling Using the Ground, First Edition.
Marc A. Rosen and Seama Koohi-Fayegh.
© 2017 John Wiley & Sons, Ltd. Published 2017 by John Wiley & Sons, Ltd.

energy systems that utilize the energy of the ground. The coverage includes details on applications and benefits of using heat pumps with short-, medium- and long-term thermal storage, for heating and/or cooling applications.

4.2   Thermal Energy Storage Methods

4.2.1   Fundamentals

TES is the storage of thermal energy (heat or cold) for a period of time in a storage medium. Thermal energy may be stored by elevating or lowering the temperature of a medium (i.e., altering its sensible heat), by changing the phase of a medium (i.e., altering its latent heat), and by causing a medium to undergo endothermic and exothermic chemical reactions. Consequently, there exist three main TES types: sensible, latent, and thermochemical.

Typical storage materials for sensible storage include water, rock, and soil, for latent storage include water/ice and salt hydrates, and for thermochemical storage include various reacting pairs of chemicals, with one sometimes being water. Sensible and latent TES are relatively mature, while thermochemical energy storage is relatively new and in development. TES in such media as aquifers, boreholes, phase change materials (PCMs), and thermochemical substances have been widely described (Paksov 2007; Dincer and Rosen 2011).

A TES system generally consists of a storage medium, a container, and equipment for injecting and recovering thermal energy. The container retains the storage material and prevents losses of thermal energy. The container can be an artificially created device or a naturally occurring structure or region.

TES has many practical applications. TES helps offset the mismatch between periods when thermal energy (heat or cold) is available and in demand. TES offers the possibility of storing thermal energy for later use in its original form, or in conversion to electricity or other energy products. TES is often used with heating and cooling technologies. Examples of TES are the storage of solar energy for overnight heating, of summer heat for winter use, of winter ice for space cooling in summer, and of the heat or cool generated electrically during off-peak hours for use during subsequent peak demand hours. Thus, TES can help efforts for demand side management (Arteconi *et al.* 2012).

District heating and cooling systems often also incorporate TES and can benefit from its careful integration into the overall system (Sibbitt *et al.* 2012; Allegrini *et al.* 2015; Rezaie *et al.* 2015).

The design or selection of a TES system mainly depends on such factors as the application, the storage period required, economics, and operating conditions.

Significant advances achieved in TES over the last couple of decades are incorporated throughout this chapter. Extensive research on TES has been undertaken through the Energy Conservation through Energy Storage Implementing Agreement (ECESIA) of the International Energy Agency (www.iea-eces.org), the Smart Net-zero Energy Building Research Network in Canada (SNEBRN 2015) which is dedicated to developing concepts and designs for smart net-zero energy buildings and communities often including TES, as well as via other research programs. In addition, this chapter draws heavily on material in Chapter 8 in Section 3 of the *Alternative Energy and Shale Gas Encyclopedia*

(Rosen 2015) and Chapter 3 of *Thermal Energy Storage: Systems and Applications* (Dincer and Rosen 2011).

4.2.2 Advantages of Thermal Energy Storage

TES systems can yield significant benefits, which vary by application (Beckman and Gilli 1984; Ataer 2006; ASHRAE 2007; Dincer and Rosen 2011; Desgrosseilliers *et al.*, 2013; Dickinson *et al.*, 2013; Rad *et al.*, 2013; Eslami-nejad and Bernier, 2013; Koohi-Fayegh and Rosen, 2014). Many of the benefits are summarized in Figure 4.1.

Some of the benefits of TES related to technical performance are:

- **Improved operation of systems in which thermal storage is applied:** The incorporation of TES into a system (e.g., heat pumps, power plants, cogeneration plants) can improve the operation of system. For instance, TES facilitates improved operation of thermal equipment, in that it can allow such equipment to operate more effectively and flexibly. Also, TES systems can complement heat pumps for heating or cooling by providing hot or cold reservoirs, thereby improving their efficiencies and performances. Also, a cogeneration plant with TES need not follow a thermal load and can be operated more advantageously. This helps overcome a weakness of cogeneration plants, in that they are generally operated to meet the demands of the connected thermal load, which often results in excess electric generation during periods of low electricity demand.
- **Load shifting:** TES permits energy consumers to shift energy use from high- to low-demand periods. Also, since demands for heating, cooling, or electricity are seldom constant over time, the excess generation capacity available during low-demand

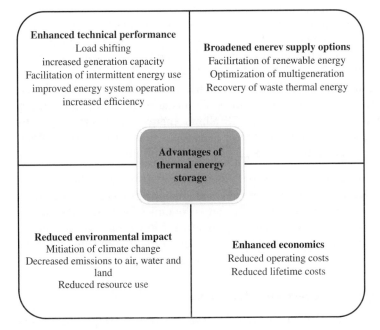

Figure 4.1 Various advantages of utilizing thermal energy storage in energy systems compared with energy systems without thermal energy storage.

periods can be used to charge TES in order to increase the effective generation capacity during high-demand periods. This benefit allows smaller production units to be installed, or increased capacity to be attained without purchasing additional units.

- **Increased efficiency:** By storing heat (e.g., waste heat, solar energy) or cold so that it can be used when needed, with temperature enhancement where necessary via heat pumps or other technologies, the efficiencies of heating and cooling operations can be increased.
- **Improved operation of thermal equipment:** TES can allow thermal equipment to operate more effectively and flexibly.
- **Facilitation of use of intermittent energy sources:** TES can facilitate the use of energy sources which are not available continuously, by storing energy between periods of availability and demand. Intermittent energy sources include waste heat, cogenerated heat and renewable energies such as solar and wind. TES thereby allows intermittent energy sources to meet a greater fraction of the loads for which they are used.

Other benefits of TES relate to environmental and economic performance, and are as follows:

- **Facilitation of use of renewable energy resources:** TES can facilitate the use of renewable energy sources, many of which are not available continuously, by storing energy between periods of availability and demand. Such intermittent renewable energy sources include solar and wind energy. TES thereby allows renewable energy sources to meet a greater fraction of the loads and facilitates substitutions of renewable energy resources at small, intermediate and large scales.
- **Reduced environmental impact:** By increasing the efficiency of systems that utilize TES and facilitating the use of renewable energy sources and waste energy, TES systems help reduce emissions of pollutants and environmental impacts. Climate change mitigation with TES has been examined (Paksoy 2007).
- **Enhanced economics:** Many of the above benefits allow TES systems to provide significant economic gains over their lifetimes. For instance, by facilitating shifting of energy use to low-demand periods, TES allows energy consumers subject to time-of-day pricing to shift energy purchases from high- to low-cost periods. Also, TES can allow thermal equipment to operate more economically.

An investigation of applications of TES for cooling capacity (Dincer and Rosen 2001) that assessed TES with exergy methods in the hope of attaining more realistic determinations of energy savings, emissions reductions and economics found that the appropriate type of TES provides energy savings of up to 50% and environmental benefits in terms of greenhouse gas (GHG) emission reductions of up to 40%. This investigation helped confirm the benefits of utilizing cold TES in energy systems.

4.2.3 Thermal Energy Storage Operation and Performance

TES systems are generally designed to operate on a cyclical basis (usually daily, weekly, or seasonally). In considering TES operation, it is useful to characterize TES systems according to storage duration:

- **Seasonal/long-term:** Long-term TES operates on annual or seasonal cycles, and usually takes advantage of seasonal climatic variations (Novo *et al.* 2010; Pinel *et al.* 2011; Pavlov & Olesen 2012; Terziotti *et al.* 2012; Guadalfajara *et al.* 2014; Yan *et al.* 2016). Seasonal TES systems have a much greater capacity than daily TES, often by two orders of magnitude. Thermal losses are more significant for long-term storage, so more effort is made to prevent thermal losses in seasonal rather than daily TES. While diurnal systems can generally be installed within a building, seasonal storage requires such large storage volumes that special care is required in locating the storage and separate locations are often required. The significance of long-term TES applications is growing in many parts of the world.
- **Weekly/medium-term:** Medium-term TES operates on weekly cycles, and exhibits many of the characteristics of short-term systems.
- **Diurnal/short-term:** Short-term (or diurnal) TES addresses peak loads lasting a few hours to a day in order to reduce the sizing of systems and/or to take advantage of energy-rate daily structures, or to allow intermittent energy sources to be used throughout the day. The use of diurnal TES for electrical load management in buildings is increasing. TES allows electricity consumption costs to be reduced by shifting electrical heating and cooling loads to periods when electricity prices are lower, usually during the night. Load shifting can also reduce demand charges, which can represent a significant proportion of total electricity costs for commercial buildings.

In cold TES applications, several strategies are available for charging and discharging so as to meet cooling demand during peak hours. There are two main operating strategies:

- **Full storage:** A full-storage strategy shifts the entire peak cooling load to off-peak hours, a strategy that is most attractive when peak demand charges are high or the peak period is short.
- **Partial storage:** With partial storage, the chiller operates to meet part of the peak-period cooling load, and the rest is met by drawing from storage. Partial-storage systems are therefore load leveling and demand limiting.

Numerous examples of TES systems exist. A large cold TES is currently being constructed in Chicago, USA to supply two-thirds of the peak-load air-conditioning demands of the downtown area. In another large development in Shanghai, China, where the electrical peak load is a serious problem and air-conditioning is responsible for a large part of the summer peak load, Hangchow City adopted an incentive rate system for buildings that adopt certain cold TES options (Saito 2002).

TES performance is affected by many factors, as illustrated in Figure 4.2. These factors include design, temperature, storage operation container and balance of system, and the setting in which the storage operates.

4.2.4 Thermal Energy Storage Types

The three main types of TES – sensible, latent, and thermochemical – are described in this section. Energy is stored or discharged by changing the temperature of the storage material in sensible storage, changing the phase of the storage material in latent storage, and changing the chemical form of the storage material in thermochemical storage.

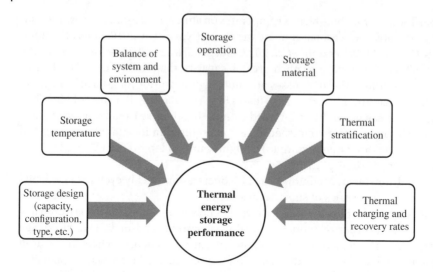

Figure 4.2 Various factors that determine the performance of a thermal energy storage unit.

4.2.4.1 Sensible Thermal Energy Storage

Sensible TES systems undergo changes in sensible heat, associated with temperature change. Some characteristics preferred for a sensible heat storage medium are: high specific heat capacity; long-term stability under repeated charging and discharging cycles; compatibility between storage medium and its containment; and low cost. Sensible heat storage systems are sometimes classified based on heat storage media as liquid (e.g., water, oil) and solid (e.g., rock, ground).

Some common types of sensible TES are summarized in Figure 4.3 and have been described extensively (Beckman and Gilli 1984; Dinter *et al.* 1991; Ataer 2006; IEA 2010; Dincer and Rosen 2011). Many sensible TES systems involve underground thermal storage, as explained in greater detail in the next main section.

Two of the more notable sensible TES types are as follows:

- **Underground:** There are many types of underground TES. Heat or cold can be transferred to soil for storage and later recovery. Ground-based storages can be shallow (e.g., Earth beds) or deep (e.g., systems consisting of boreholes in the ground that act as ground heat exchangers for transferring heat or cold into underground soil and rock). The use of the ground as a storage medium is often simpler for new construction; the required below ground work can make retrofits difficult, especially if work is needed beneath a structure. Also, an aquifer can be used as a type of underground storage, by extracting water from one location in the aquifer for heating (or cooling) and then reinjecting it at another point in the aquifer for storage. Finally, a salinity gradient solar pond can at an integrated collection and below ground storage device for solar energy.
- **Container/tank:** Containers and tanks filled with a heat storage medium, such as water or rock, can act as a TES. Container/tank-based storage can be located in or above ground. Such tanks are often made of steel or concrete because of their physical characteristics, cost, availability, and easy processing. Ceramic bricks can also act as a good heat storage medium, especially for uses in new and old buildings,

Figure 4.3 Various types of sensible thermal energy storage. Ground-based storages are separated into shallow systems, which mainly utilize soil as a storage medium, and deep systems, which rely predominantly on rock as the storage medium.

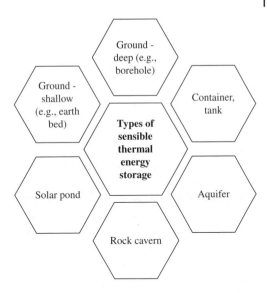

where they are advantageous due to their modular sizes, ease of installation, and high heat-retention abilities. The container can be artificial or natural. An example of the latter case is the use of a rock cavern filled with a storage medium as TES.

It is useful to know the system characteristics for the two main types of TES (underground and container/tank), so their characteristics are contrasted in Table 4.1. The sizes can exceed $500\,000\,m^3$, while the energy efficiencies typically vary from 60 to 90% for short-term applications, and less for longer term applications. A storage's specific thermal capacity (i.e., storage capacity per unit temperature rise on a per unit volume basis) is useful for comparing its thermal storage capacity with others.

The thermophysical and economic characteristics of sensible thermal storage materials have been recently investigated (Fernandez *et al.* 2010). Sensible heat storages commonly use water or ground or rock as the storage medium due to their favorable characteristics: widespread availability, low cost, ease of use and appropriate thermal behavior. Water's high heat capacity often makes water tanks suitable for TES systems for building heating or cooling. The relatively low heat capacities of rock and ceramics are somewhat offset by the large temperature changes possible with these materials, and their relatively high densities. Hasnain *et al.* (1996) propose cast iron as a sensible

Table 4.1 Characteristics for two main types of sensible thermal energy storage (TES).

Characteristic	Type of sensible TES	
	Underground (borehole, earth bed, rock cavern, aquifer)	Tank/container (steel, concrete)
Volume (typical upper value) (m^3)	500 000	100 000
Specific thermal capacity $(kJ/m^3\,K)$	2270–4180	4180
Energy efficiency (for short-term applications) (%)	60–80	90

Data sources: SFEO (1990) and Piette (1990).

heat storage material, noting its energy density exceeds that for water, but that its cost exceeds that for rock or brick. Pebble beds or rock piles are sometimes advantageous choices for sensible storage.

Sensible TES can accommodate a wide range of storage durations. Short-term (daily to weekly) systems often use crushed rock beds, earth beds, water tanks, ceramic brick, and building mass. Long-term (annual or seasonal) TES systems often utilize rock beds, earth beds, and water tanks. Aquifer TES can be used for all storage periods.

Some sensible TES systems exploit thermal stratification, in the form of a temperature gradient – usually vertical – within the storage. Thermal stratification is desirable as it avoids loss of temperature due to mixing, and is easier to achieve in solid storage media than in fluids.

Sensible heat storage is commonly applied with solar water heaters, to store solar thermal energy for periods without sunshine. The use of packed bed TES with solar air heaters has advanced recently in various facets (Singh *et al.* 2010): storage material and design, void fraction, heat transfer, air flow pattern, and pressure drop. The materials considered most are rocks and pebbles, and small to large packed beds are common. Sensible TES systems are often long-term (particularly annual or seasonal) units, and their use is expanding in many parts of the world. A noteworthy example, as described in detail later in this chapter, is the Drake Landing Solar Community (DLSC) in the city of Okotoks, Alberta (SAIC 2010; Dincer and Rosen 2011). The project successfully integrates long- and short-term sensible TES technologies with solar energy. The corresponding district heating system is designed to store solar energy underground in summer months and supply it to the 52 homes of the DLSC community in winter months. This project meets about 90% of the space heating requirements of the community.

4.2.4.2 Latent Thermal Energy Storage

Latent heat changes are the heat interactions during the phase change of a material at constant temperature, for example, solid from/to liquid (Fang *et al.* 2010; Cabeza *et al.* 2011). Latent heat TES systems store or release thermal energy as a material changes phase. A latent heat change is usually much higher than a sensible heat change for a given amount of storage medium, giving latent TES a higher energy storage density. Latent heat TES systems incorporate a storage containment and capability for transferring thermal energy into and out of the storage.

Latent heat TES systems utilize storage media, usually selected to undergo phase change within the desired operating temperature range. Phase change materials include inorganic and organic materials, large fatty acids, aromatics salt compounds that absorb a large amount of heat during melting (e.g., eutectic salts, salt hydrates, Glauber's salt), and paraffin waxes. The latter are common because they have good stability and do not degrade notably over repeated storage cycles.

The phase change material in a latent TES can be contained in a single large vessel or in small modules (e.g., rods, plastic containers). The use of small modules provides great flexibility in latent TES applications.

Latent heat TES can be used to store hot or cold. Cold latent TES typically uses the following storage media: water, eutectic salts, glycol, brine, and ice slurry.

Phase-change materials (PCMs) provide greater energy storage capacity with smaller temperature fluctuations compared with sensible storage. PCMs also are potentially advantageous for use in solar walls (Farid and Kong 2001; Farid *et al.* 2004; Zalba *et al.* 2004). Khalifa and Abbas (2009) examined thermal storage walls with three PCMs [concrete, hydrated salt ($CaCl_2 \cdot 6H_2O$) and paraffin wax (N-eicosane)], and their abilities to maintain a comfortable space temperature. Wall thickness and heat storage media were observed to significantly affect temperature control, and a storage wall of hydrated salt (thickness 0.08 m) was demonstrated to be capable of maintaining a space temperature of $18-22\,°C$.

Jegadheeswaran and Pohekar (2009) reviewed various techniques to improve the heat transfer rates in PCMs, and shorten charging and discharging periods. Recommended measures include the use of extended surfaces and multiple PCM configurations, thermal conductivity enhancements, micro-encapsulation PCMs, and the use of exergy methods to identify additional areas of potential improvement. Work is also ongoing to enhance PCM thermal conductivity, by developing appropriate material combinations, using metal matrixes and high conductivity particles in materials, improving fin configurations, and applying bubble agitation and micro-encapsulation (Cabeza *et al.* 2007; Shilei *et al.* 2007; Alkan and Sari 2008; Koca *et al.* 2008; Sari and Karaipekli 2009; Sharma *et al.* 2009;Agyenim *et al.* 2010; Bayés-García *et al.* 2010; Lai *et al.* 2010; Fan and Khodadadi 2011). Heat transfer advantages have been demonstrated by using carbon-fiber chips and carbon brushes as PCM additives (Hamada *et al.* 2003) and paraffin with Lessing rings (Velraj *et al.* 1999). A coated shell and tube heat exchanger using a specific salt mixture composition was shown effective for avoiding the heat transfer rate reduction associated with PCM solidification around heat exchanger tubes (Mathur *et al.* 2010).

In solar combi-systems for space heating and cooling, large solar factions normally require larger water volumes, but PCMs offer more compact alternatives. Heinz and Schranzhofer (2010), through an analysis of PCM and water storage with dynamic system simulation, demonstrate that PCMs are competitive with sensible storage for a small tank volume and low solar fraction, and that PCMs can be advantageous to traditional heat storage methods for seasonal storage.

4.2.4.3 Thermochemical Thermal Energy Storage

Thermochemical storage utilizes chemical reactions to store and release heat. It is based on a chemical reaction that can be reversed:

$$C + heat \rightleftarrows A + B \tag{4.1}$$

Thermochemical material (C) absorbs energy and is converted chemically into two components (A and B), which can be stored separately. The reverse reaction occurs when A and B are combined together and C is formed, releasing the thermal energy that is recovered from the TES. The storage capacity of this system is the heat of reaction when C is formed. Substance C is the thermochemical material for the reaction and can be a hydroxide, hydrate, carbonate, or ammoniate, while A and B are reactants, which can be water, carbon monoxide, ammonia, and hydrogen. Usually material C is a solid or a liquid and A and B can be any phase.

Thermochemical storages are generally more compact than latent and sensible storages. Thermochemical TES systems are not yet commercial but have been considered for several decades, for example, Wettermark (1989) comprehensively reviewed thermochemical energy storage and criteria for choosing appropriate storage media and candidate reaction pairs. Further research is ongoing to better understand and design

Table 4.2 Principal processes in thermochemical energy storage.

Process	Reaction	Reaction type (thermal)
Charging	C + heat → A + B	Endothermic
Storing	None	None
Discharging	A + B → C + heat	Exothermic

thermochemical storages and to address other issues affecting commercial implementation (Bales 2006; Masruroh *et al.* 2006; Hauer and Lavemann 2007; IEA 2008; Zondag *et al.* 2008).

The design of thermochemical energy storage systems is complex and requires appropriate consideration of many factors. This category of TES includes sorption and thermochemical reactions.

The three main processes (charging, storing, discharging) of a general TES cycle are listed for thermochemical energy storage in Table 4.2 and described individually below:

- **Charging:** During the endothermic charging process [Equation (4.1) in forward direction], thermal energy is absorbed from an energy resource, causing the dissociation of the thermochemical material (C) into two materials (A and B).
- **Storing:** Substances A and B are stored separately, usually at ambient temperatures. Material degradation can lead to some energy loss, but there is little heat loss except during the initial cooling of components A and B after charging.
- **Discharging:** Substances A and B combine exothermically during discharging [Equation (4.1) in backward direction], allowing the stored energy to be recovered and material C to be regenerated for reuse in the cycle.

Thermal energy is stored after a dissociation reaction and then recovered in a chemically reversed reaction during thermochemical energy storage. This storage type is particularly suited to long-term storage applications such as seasonal storage of solar heat because it typically incurs small energy loss during the storing period as it usually is at ambient temperature. In sorption systems (adsorption and absorption), adsorption occurs when an adsorptive (usually liquid or gas) accumulates on the surface of an adsorbent usually (solid or liquid) and forms a molecular or atomic layer. In absorption, the substance is distributed into a liquid or solid and forms a solution.

Several investigations of thermochemical storage materials and their characteristics have been undertaken. Long-term heat storage for a closed sorption system using the working pair NaOH and water has been contrasted with conventional storage (Weber and Dorer 2008). An examination of the potentials of $MgCl_2 \cdot 6H_2O$, $CaCl_2 \cdot 2H_2O$, $MgSO_4 \cdot 7H_2O$, and $Al_2(SO_4)_3 \cdot 18H_2O$ as salt hydrates for thermochemical storage showed that each can be dehydrated below 150 °C; the hydration and dehydration behavior of various $MgCl_2 \cdot 6H_2O$ was the most promising (van Essen *et al.* 2010). Thermochemical storage has been studied based on strontium bromide as the reactant and water as the working fluid (Lahmidi *et al.* 2006). The working pair $MgSO_4$ and Zeolite have been shown to have reasonable energy densities and thermal properties for seasonal thermochemical heat storage (Hongois *et al.* 2010). The working pair strontium bromide and water have been analyzed experimentally for heating and

Table 4.3 Candidate materials for thermochemical energy storage, broken down by thermochemical reaction basis.

Basis of thermo-chemical reaction	Solid reactant (A)	Working fluid	Thermochemical storage material (C)	Thermophysical characteristics	
		(B)		Energy storage density (GJ/m^3)	Charging reaction temperature (°C)
Water-based	$MgSO_4$	H_2O	$MgSO_4 \cdot 7H_2O$	2.8	122
	CaO	H_2O	$Ca(OH)_2$	1.9	479
	FeO	H_2O	$Fe(OH)_2$	2.2	150
	$CaSO_4$	H_2O	$CaSO_4 \cdot 2H_2O$	1.4	89
Carbon dioxide-based	FeO	CO_2	$FeCO_3$	2.6	180
	CaO	CO_2	$CaCO_3$	3.3	837

Source: Adapted from Visscher and Veldhuid (2005).

cooling (Mauran *et al.* 2008). The thermochemical storage reaction pair ammonia and water have been examined for solar thermal applications, including the characteristics of the materials and their dissociation (Kreetz and Lovergrove 1999; Lovergrove *et al.* 1999a, b). The cycling behavior of the storage medium magnesium sulphate ($MgSO_4$) and the dehydration temperature of the reactant have been investigated (Zondag *et al.* 2007). The reversibility and efficiency of thermochemical storage materials $Ca(OH)_2$ and CaO have been investigated (Azpiazu *et al.* 2003).

Selected thermochemical storage material candidates are listed in Table 4.3. They are based on water and carbon dioxide working fluids. Also listed in Table 4.3 are energy densities and reaction temperatures for the thermochemical material. These are important thermal factors that strongly affect their application in thermochemical TES systems.

Several investigations of thermochemical storage applications have been reported. Exergy analysis has recently been applied to thermochemical storage (Haji Abedin and Rosen 2010a, b). For a direct floor heating system using flat plate solar collectors, the relations have been determined between power levels and heating storage capacities for a thermochemical storage using strontium bromide as the reactant and water as the working fluid (Lahmidi *et al.* 2006).

4.2.5 Thermal Energy Quality and Thermal Energy Storage Stratification

4.2.5.1 Thermal Energy Quality

Energy quality is represented by the temperatures of the materials entering, leaving and stored in a storage, and is an important storage consideration. For example, 1 MWh can be stored by heating 1000 Mg of water 0.86 °C, or by heating 10 Mg of water 86 °C. The latter case is more advantageous in terms of energy quality as a wider range of tasks can be accomplished with the higher temperature medium upon discharging. The costs are usually higher for thermal storages that retain thermal energy quality, that is, temperature.

Table 4.4 Factors to be considered in designing or selecting a thermal energy storage system for a given application, divided by category.

Requirements of thermal energy storage	Performance measures	System-related limitations	Economic parameters	Environmental considerations
• Storage capacity • Storage duration • Charging rate • Discharging rate • Variability in thermal loads	• Energy efficiency • Exergy efficiency • Reliability • Impact on application • Health and safety • Lifetime	• Size (volume) • Surface space (area) • Installation limitations	• Initial capital cost • Annual operating cost • Amortized cost • Payback period	• Environmental emissions associated with storage • Resource consumption in installing and operating storage • Change in environmental impact of application due to storage

4.2.5.2 Exergy of Thermal Energy

Exergy provides a useful measure of energy quality and can be beneficially used for thermal storage analysis and improvement. Exergy efficiency is a useful performance factor (Table 4.4); it is akin to the more common energy efficiency but significantly different (Rosen *et al.* 2004; Dincer and Rosen 2007).

Exergy can informally be viewed as a measure of energy quality or usefulness or value, and it is formally defined as the maximum amount of work which can be produced by a stream or system as it comes into equilibrium with a reference environment. Exergy is thus evaluated relative to a reference environment. Exergy is consumed during real processes due to irreversibilities, and conserved during ideal processes.

Exergy analysis as a tool has been applied to various processes (Dincer and Rosen 2007), including thermal energy storage (Rosen *et al.* 2004; Dincer and Rosen 2011). Exergy analysis involves the evaluation of exergy efficiencies, which – unlike energy efficiencies – always measure how nearly process operation approaches the ideal or theoretical upper limit. Exergy inefficiencies are in a process that represent the largest theoretical margins for efficiency improvement. Consequently, exergy analysis is beneficial for TES systems, be they for heat and cold because it allows meaningful efficiencies to be evaluated and sources of thermodynamic losses to be clearly identified and quantified (Dincer and Rosen 2011).

4.2.5.3 Storage of Thermal Exergy

It is useful to consider the storage of thermal energy at useful temperatures as TES. An illustration can help demonstrate this point and show how exergy efficiencies provide rational performance measures relative to an ideal.

Consider a perfectly insulated TES containing 1 Mg of water, initially at 40 °C. The ambient temperature is 20 °C, and the specific heat of water is assumed fixed at 4.2 kJ/kg K. Heat (4200 MJ) is transferred to the storage through a heat exchanger from an external body of 0.1 Mg of water cooling from 100 to 90 °C. This heat addition raises the storage temperature by 1.0 °C, to 41 °C. After a storage period, 4200 MJ of heat is recovered from the storage through a heat exchanger which delivers it to an external body of 0.1 Mg of water, raising its temperature from 20 to 30 °C and returning the storage to its initial temperature of 40 °C.

For this storage cycle the energy efficiency is the ratio of the heat recovered from the storage to the heat injected, that is, 4200 MJ/4200 MJ = 1 (or 100%). The recovered heat is of little use being at only 30 °C; it was degraded even though the storage energy efficiency is 100% so no heat was lost. The exergy recovered in this example is 0.07 MJ and the exergy supplied is 0.856 MJ, making the exergy efficiency 0.07/0.856 = 0.082 (8.2%). Being the ratio of the thermal exergy recovered from storage to that injected, the exergy efficiency provides a meaningful measure of the achieved performance of the TES. It also shows the practical and theoretical margin for efficiency improvement. That is, a device that seems ideal on an energy basis is correctly shown to be far from ideal on an exergy basis, clearly demonstrating the benefits of using exergy analysis on TES.

4.2.5.4 Thermal Stratification

Exergy analysis has been shown to be useful for quantifying thermal stratification and its advantages (Rosen *et al.* 2004). Haller *et al.* (2009) suggest methods for characterizing thermal stratification efficiency in thermal storage systems and report stratification efficiency to be a significant parameter. Existing methods are used to evaluate stratification efficiencies for the charging, discharging, storing and overall processes, and compared with the entropy production rate due to mixing. The stratification efficiency is taken to be 0% for a fully mixed tank and 100% for a perfectly stratified tank.

4.2.6 Thermal Energy Storage Economics

TES systems often provide energy cost savings relative to the primary generating equipment required to satisfy the same service (loads, periods, etc.), but often with higher initial capital costs. The economics of TES systems normally requires that their annualized capital and operating costs not exceed the costs of the primary equipment.

The economics of thermal storage is usually evaluated considering numerous factors (see summary in Figure 4.4), including:

- temporal variations of heating, cooling and electrical loads;
- storage system design characteristics (e.g., type, size);
- storage operation characteristics (e.g., efficiency, control);
- costs (e.g., electricity consumption costs, time-of-use cost adjustments, electricity demand charges, TES costs);
- other economic factors (e.g., financial incentives).

Economic analyses of TES systems are important to their design, particularly since thermal storages sometimes have relatively high initial capital costs which are balanced against savings over time. Economic evaluation criteria often include simple payback period and return on investment. Payback periods often vary significantly with application, for example, from 1 to 10 years in many instances for cold TES systems. TES systems can be applied cost-effectively in many residential, institutional and commercial buildings, often exhibiting payback periods under 3 years. The Electric Power Research

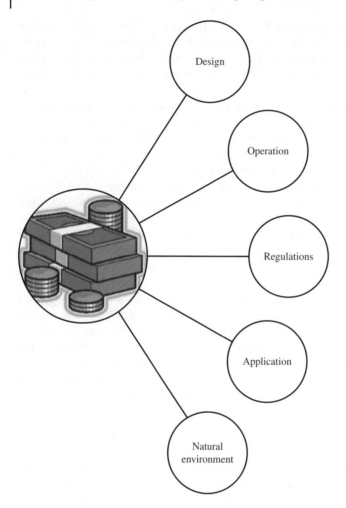

Figure 4.4 Various factors that affect the economics of thermal energy storage systems.

Institute estimates that overall HVAC operating costs can be reduced by as much as 60% with thermal storage. Economic comparisons often consider whether annual energy cost savings can exceed the annualized capital cost. TES systems for cooling capacity are typically more advantageous in larger buildings, although the benefits of smaller applications have been increasing.

Thermal storage applications in buildings also provide benefits to electricity utilities by shifting electrical loads to off-peak periods. The resulting reduction in peak electrical demand reduces the stresses on generating equipment, lowers the amount of generation capacity required, and curtails the use of peaking plants, which typically operate more expensively and with greater emissions than baseline generating stations. For this reason, electrical utilities often provide incentives for the use of TES, and thermal storage economic viability is often significantly affected by such government and utility financial incentives, for example, special electricity rates, rebates, and tax reductions.

4.2.6.1 Economics of Thermal Storage for Cooling

TES permits energy cost savings by allowing the capacity of cooling systems to be reduced, often by over half. This can be due to better energy management and the attainment of higher system efficiencies.

In particular, TES systems permit consumer electricity costs for space cooling to be lowered by shifting electrical loads to periods of lower electricity prices under time-of-use pricing. Load shifting can also reduce demand charges.

Cooling TES systems are generally justified for new facilities which have cooling loads that are large during peak periods (e.g., days).

Cooling thermal storage is often not advantageous in retrofit situations, except where the cooling system is being replaced for other reasons (inadequate capacity, old age, etc.).

4.2.6.2 Economics of Thermal Storage for Heating

Heating thermal storage systems can often be justified economically for heated facilities with large space heating needs that are temporally mismatched with a thermal supply (e.g., solar thermal energy).

Heating thermal storage systems also can often be advantageous economically for electrically heated facilities with large space heating needs that are charged under time-of-use electric rates. This is particularly true when the difference between peak and off-peak electric rates is large, and/or when electricity use without TES mainly occurs during peak periods. Shifting the heating load can also reduce electrical demand charges.

Thermal storage systems can be employed to shift any portion of the energy use for space heating to off-peak periods. This contrasts notably with the energy use for heating in off-peak periods for conventional space heating systems, which is typically closer to half of the energy use for heating.

Thermal storage often permits energy cost savings by allowing the capacity of heating systems to be reduced, sometimes by over 50%, through better energy management and higher efficiencies.

4.2.7 Thermal Energy Storage Design, Selection, and Testing

Energy demands in the commercial, industrial, utility and residential sectors for such tasks as space and water heating, cooling and air conditioning, vary on daily, weekly and seasonal bases. A TES is designed or selected to match the application. Numerous other criteria also affect the selection of TES systems, many of which are summarized in Table 4.4. These criteria include technical factors (e.g., required storage capacity, storage duration, physical size, space availability, efficiency, installation limitations, reliability, safety, impact on performance of the overall application), environmental factors, and economics (e.g., system cost, lifetime, payback period). Appropriate trade-offs are often made among competing criteria. The Air Conditioning Contractors of America (ACCA) has developed a guide to explain TES using HVAC terminology, aimed at providing designers and contractors with a step-by-step approach. The guide aims to increase understanding of TES technologies and their integration with applications, and to elevate comfort levels so that contractors can consider TES options more readily (ACCA 2005; ASHRAE 2007).

Fernandez *et al.* (2010) developed a methodology for the selection of suitable sensible heat storage materials in the temperature range 150–200 °C. In this approach the material selection was based on volumetric energy density as well as cost and thermal conductivity (preferably greater than 1 W/m K). The approach was reported to be helpful in the selection of suitable TESs, basis costs, availability, and environmental benefits.

The enhancement of thermal conductivity is the leading area of research in PCMs. In recent years, considerable efforts have been made to develop appropriate techniques and/or combinations of materials to increase PCM thermal conductivity (Cabeza *et al.* 2007; Sari and Karaipekli 2009; Agyenim *et al.* 2010). Some of the recommended measures include finned tubes of various configurations, bubble agitation, insertion of a metal matrix into the PCM, using PCMs dispersed with high conductivity particles, micro-encapsulation, and multi-tube configurations. Hamada *et al.* (2003) used carbon-fiber chips and carbon brushes as additives and observed the carbon-fiber chips to be effective in improving the heat transfer rate in PCMs. However, the higher thermal resistance near the heat transfer surface was observed to lower the overall heat transfer rate in the fiber chips relative to carbon brushes, indicating that the carbon brushes are superior to fiber chips. Other studies have also reported enhancements in heat transfer rate for innovative measures (Alkan and Sari 2008; Bayés-García *et al.* 2010; Lai *et al.* 2010; Fan and Khodadadi 2011). One significant development was reported by Velraj *et al.* (1999) who investigated paraffin and Lessing rings, and found this combination improved the thermal conductivity about 10 times compared with paraffin (i.e., to 2 from 0.2 W/m K).

A constraint preventing the large-scale commercial use of PCMs for energy storage is salt solidification around the heat exchanger tubes, which slows the discharging of the latent heat stored in PCMs. Terrafore has developed an innovative approach that helps to increase the heat transfer rate by forced convection (Mathur *et al.* 2010). In this technique a coated shell and tube heat exchanger is used with a specific salt mixture composition (i.e., a dilute eutectic). This heat exchanger with a unique coating of salt mixture has been observed to improve the heat transfer coefficient, and work is continuing to further develop it.

An alternative method of numerical modeling for designing TES systems with PCMs has been investigated by Bruno *et al.* (2010). The method, based on the effectiveness-Number of Transfer Units (ε-NTU) approach, has been tested experimentally on a cylindrical tank filled with PCM. The results show that the ε-NTU technique provides a useful design tool for the sizing and optimizing of a TES unit with PCMs.

Several approaches exist for testing TES systems (Scalat *et al.* 1996; Hill *et al.* 1977; Henze *et al.* 2003). Some examples are:

- ASHRAE and ANSI provide testing standards for some types of TES (ASHRAE 2000). These include ASHRAE standard 943 (Method of Testing Active Sensible TES Devices Based on Thermal Performance) and ANSI/ASHRAE standard 94.1-1985 (Method of Testing Active Latent Heat Storage Devices Based on Thermal Performance). Note that these were originally combined in ASHRAE TES standard 94-77: Methods of Testing Thermal Storage Devices Based on Thermal Performance.
- The Canadian Standards Association published information on the design and installation of underground TES systems for commercial and institutional buildings (CSA 2002), which includes information on testing.

It is noted that the complexity of the processes involved in TES has made the development of reliable test methodologies that predict performance challenging.

National and international standards for the testing of thermal storage using PCMs are lacking (Agyenim *et al.* 2010). A similar situation is true for thermochemical storage. This is despite the fact that research is ongoing to develop PCMs and thermochemical storage materials for various applications. Unified standards for thermal storage testing and performance comparison are needed to foster growth in applications but efforts are being expended to achieve them.

4.2.8 Thermal Energy Storage Markets and Applications

4.2.8.1 Thermal Energy Storage Markets

TES applications have achieved various levels of market penetration, depending on the country. Diurnal heat storage has achieved a large market share in many countries. Diurnal cold storage in air conditioned buildings for demand-side management is growing. Short-term cool TES for air conditioning is often cost-effective, with numerous applications in the USA, Canada, Japan, and Europe. Individual houses or commercial buildings can use diurnal TES either for heating or cooling applications (Heim 2010; Novo *et al.* 2010; Pinel *et al.* 2011; Pavlov and Olesen 2012; Terziotti *et al.* 2012; Tatsidjodoung and Le Pierrès 2013; Guadalfajara *et al.* 2014; Heier *et al.* 2015; Yan *et al.* 2016). Seasonal TES has achieved the greatest success in northern countries, sometimes in conjunction with district heating. Large seasonal TES has often been installed with solar collection systems in large buildings or in association with district heating.

4.2.8.2 Thermal Energy Storage Applications

Numerous applications are possible for TES. The most common areas in which TES is employed are the building, industry, and utility (e.g., power generation) sectors. Various applications of TES systems are shown in Figure 4.5.

Cold TES has many applications, for example, space conditioning in buildings and food processing. Thermal storages for cooling capacity have been examined to determine energy savings, GHG emission reductions and economic benefits (Dincer and Rosen 2001).

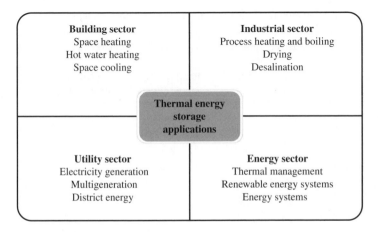

Figure 4.5 Various common applications of thermal energy storage, broken down by sector.

Much research has been reported on using natural and renewable energy resources with TES in heating and cooling applications (Paksoy 2007). The integration of TES with solar energy has significant potential. The potential of solar thermal walls is one interesting passive solar application. Some building materials such as gypsum wallboards are suitable for PCM containment, allowing PCMs to be integrated into buildings. Also, solar combi-systems (i.e., systems that provide both space heating and cooling) offer another beneficial application for TES. Various solar dryers using thermal storage have been developed although their commercial and economic success has been limited. TES permits solar dryers to extend the drying period, reduce product waste and costs, and improve drying quality. TES can be used in the continuous drying of agricultural food products at moderate temperatures (40–75 °C) using solar energy (Bal *et al.* 2010), and latent storage can be advantageous to sensible storage for this application since it keeps the drying air temperature nearly constant.

TES is a critical component in efforts to develop net-zero energy buildings and communities, that is, buildings and communities that achieve zero average annual energy consumption at both the building and community levels. Such buildings and communities will likely utilize short-term and seasonal TES in conjunction with building-integrated solar energy systems, high performance windows with active control of solar gains, heat pumps, combined heat and power technologies, and smart controls.

4.2.9 Comparison of Thermal Energy Storage Types

In Table 4.5, selected thermophysical properties of several sensible and latent TES materials are provided. For these TES materials, the storage size, in terms of mass and volume, needed to store a fixed quantity of thermal energy is also listed in Table 4.5.

The storage masses and volumes needed are observed to exhibit significant variabilities. It can be seen that the storage having the lowest mass density requires a volume 15 times greater than the storage having the greatest mass density, while the storage having the lowest volumetric density needs a volume 11 times greater than the storage having the greatest volumetric density.

Table 4.5 Selected characteristics of common sensible and latent thermal energy storage media.

Characteristic	Sensible thermal storage		Latent thermal storage	
	Rock	Water	Phase change material (organic)	Phase change material (inorganic)
Thermophysical property				
Density (kg/m^3)	2340	1000	800	1600
Specific heat (kJ/kg)	1.0	4.2	2.0	2.0
Latent heat of fusion (kJ/kg)			190	230
Storage size to store 1 GJ of energy				
Mass (Mg)	67	16	5.3	4.4
Volume (m^3)	30	16	6.6	2.7

Source: Data drawn from several sources (Hasnain 1998; Dincer and Rosen 2011).

Table 4.6 Performance factors for the three main types of thermal energy storage.

Performance parameter	Sensible thermal energy storage	Latent thermal energy storage	Thermochemical energy storage
Storage lifetime	20 yr	100–1000 storage cycles	Unknown
Storage density (GJ/m^3)	0.03–0.2	0.3–0.5	0.4–3.0
Storage temperature range (°C)	30–400	20–80	20–200
Status	Commercial	Commercial (for limited applications)	Research and development

Source: Adapted from several sources, including http://www.preheat.org/technology, Wettermark (1989), Bakker *et al.* (2009), van Helden (2009), and Dincer and Rosen (2011).

Table 4.7 Primary advantages and disadvantages for the three main types of thermal energy storage.

Type of thermal energy storage	Advantage	Disadvantage
Sensible	• Low cost • High reliability • Ease of use • Abundant materials	• Large volume • Significant heat loss over time
Latent	• Medium storage density • Small volume	• Low heat conductivity • Corrosive materials • Significant heat loss over time
Thermochemical	• High storage density • Compact (volume) • Low heat loss • Long storage period • Long distance transport	• Relatively high cost • Technical complexity • Not yet commercial

Source: Adapted from several sources, including http://www.preheat.org/technology, Wettermark (1989), Bakker *et al.* (2009), van Helden (2009), and Dincer and Rosen (2011).

Sensible, latent and thermochemical TES types are compared in Table 4.6, considering a range of performance factors. The primary advantages and disadvantages of these three storage types are compared in Table 4.7.

4.3 Underground Thermal Storage Methods and Systems

Underground TES systems are normally sensible storages, and can store both hot and cold energy (Sanner and Chant 1992; Sanner *et al.* 2003; Sharma *et al.* 2009; Singh *et al.* 2010; Cui *et al.* 2015; Dominkovic *et al.* 2015; Flynn and Siren 2015; Giordano *et al.* 2016).

4.3.1   Types and Characteristics of Underground Thermal Energy Storage

Some of the more common types of underground TES are described below (Beckman and Gilli 1984; Dinter *et al.* 1991; Ataer 2006; IEA 2010; Dincer and Rosen 2011):

- **Soil and earth bed:** Heat or cold can be transferred into underground soil for storage and subsequent recovery. Soil and earth bed TES systems are often shallow (e.g., earth beds). The use of earth as a TES medium is often restricted to new construction, since the application requires installations in the ground, beneath a structure making retrofit work difficult.
- **Borehole:** Borehole TES systems are often deep (up to hundreds of meters), consisting of a network of tubes inserted into boreholes drilled into the ground. These permit heat or cold to be transferred into underground soil and rock for storage and subsequent recovery. Borehole heat exchangers are manufactured, installed and operated in a relatively standardized manner today, and two are illustrated in Figure 4.6.
- **Aquifer:** An aquifer is a groundwater reservoir, in which the water is located in impermeable materials such as clay or rock and moves very slowly. An aquifer TES is typically a permeable, water-bearing rock formation. Aquifers often have large volumes, exceeding millions of cubic meters, and consist of about 25% water. In aquifer TES, water from the aquifer is extracted, and heated or cooled. It is then reinjected at another point in the aquifer for storage and subsequent recovery (Lee 2010; Réveillèrea *et al.* 2013). With an aquifer system, therefore, two well fields are often tapped: one for cold storage and the other for heat storage. Aquifer stores are most suited to high capacity systems. External thermal energy is stored in some aquifer TES systems, while the natural groundwater temperatures are used in others.
- **Rock cavern:** A rock cavern can be filled with a storage medium and used as a TES. The storage medium in such systems depends on the ability of the cavern to hold it. Such TES systems are usually large.
- **Container/tank:** Containers and tanks filled with a heat storage medium, such as water or rock, can act as a TES. Container- and tank-based TES types are not necessarily in-ground systems, as they can be located in or above ground. However, because they are often large and because the ground provides a degree of insulation, container- and tank-based storages are often placed underground. Such tanks are often made of steel or concrete because of their physical characteristics, cost, availability, and easy processing. They can be used for new and old buildings.
- **Solar pond:** A salinity gradient solar pond is an integrated collection and storage device of solar energy. In an ordinary pond, the sun's rays heat the water which, being less dense, rises to the surface and loses heat to the atmosphere. A solar pond inhibits this phenomenon by dissolving salt into the bottom layer of the pond, making it too dense to rise to the surface, even when hot. The salt concentration increases with depth, forming a salinity gradient. Sunlight which reaches the bottom of the pond remains trapped there as thermal energy. Useful thermal energy is recovered as hot brine.

Selected system characteristics are compared in Table 4.8 for several types of underground TES. Note that the steel and concrete tank TESs in Table 4.8 can also be located above ground. All are sensible storages, and their sizes range to as large as 500 000 m³. The efficiencies vary, for short-term applications, from approximately 60 to 90%. The

Figure 4.6 Two of the many boreholes comprising the borehole thermal energy storage system at the University of Ontario Institute of Technology, Oshawa, Ontario, Canada. The composition and geology of the ground are shown. The U-tubes that act as ground heat exchangers are seen to split off a header and then to descend to the bottom of each borehole before returning to the top.

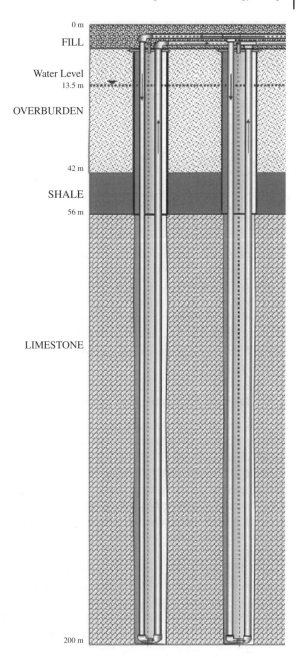

0 m

FILL

Water Level
13.5 m

OVERBURDEN

42 m

SHALE
56 m

LIMESTONE

200 m

specific thermal capacity is the storage capacity per unit volume and per unit temperature rise, and allows comparisons of storage capacity for thermal storages of the same size and same temperature increase.

Significant research is presented for on-site investigations for underground TES applications and borehole TES design using earth energy design software (Paksoy 2007).

Table 4.8 Characteristics for various types of underground thermal energy storage.

Characteristic	Ground (borehole)	Ground (Earth bed)	Steel tank	Concrete tank	Rock cavern	Aquifer
Volume (typical upper value) (m^3)	400 000	100 000	100 000	100 000	300 000	500 000
Specific thermal capacity (kJ/m^3 K)	2270	2520	4180	4180	4180	2700
Energy efficiency (for short-term applications) (%)	70	60	90	90	80	75

Source: SFEO (1990) and Piette (1990).

Numerous applications exist of underground TES applications. For instance, large-scale underground thermal storage systems are utilized at Stockton College in New Jersey, USA and in Sweden (IEA 2009). In Canada, systems are used at the Scarborough Centre in Toronto, Carleton University in Ottawa, the Sussex Hospital in New Brunswick, the Pacific Agricultural Centre in Agassiz, British Columbia and the University of Ontario Institute of Technology in Oshawa, Ontario (IEA 2009).

Underground TES is often a component in net-zero energy buildings and communities. Such buildings and communities often require seasonal TES with solar thermal collectors.

4.3.2 Example: Residential Heating Using Underground Thermal Energy Storage

In 2006, the DLSC consisting of 52 low-rise detached homes was completed in Okotoks, Alberta, Canada (McClenahan *et al.* 2006; Wong *et al.* 2006; Sibbitt *et al.* 2007). The DLSC energy system (Figure 4.7) demonstrates the feasibility of replacing substantial residential conventional fuel energy use with solar energy, collected during the summer and utilized for space heating during the following winter, in conjunction with seasonal TES. Each DLSC house has a detached garage at the back facing a lane, and the garages are joined by a roofed breezeway that extends the length of each of the four laneways and incorporates the solar collectors. The DLSC energy system has two TES types (short- and long-term).

For DLSC homes, 90% of heating and 60% of hot-water needs are designed to be met using solar energy. Each DLSC house is seen in Table 4.9 to have much lower annual energy use and GHG emissions for heating (space and domestic hot water) than a conventional Canadian home. Each DLSC home is about 30% more efficient than conventionally built houses and uses about 70% less natural gas to heat water than a conventional new home. Note that the DLSC homes have numerous other efficiency and conservation measures.

The DLSC energy system has five main components: (1) solar thermal collectors; (2) the Energy Centre, which connects the various DLSC parts and contains short-term TESs; (3) the district heating system; (4) the borehole thermal energy storage (BTES) with 144 boreholes that are 35 m deep; and (5) the solar domestic hot water system, which uses separate rooftop solar panels and natural gas-based hot-water units for winter peaking requirements.

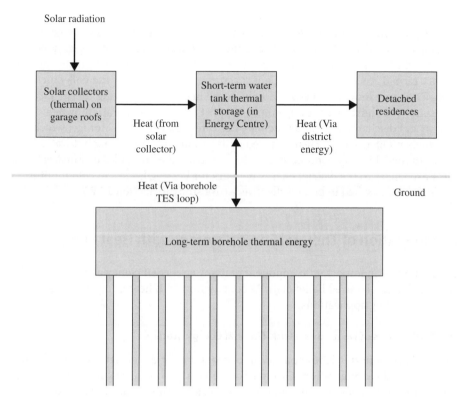

Figure 4.7 Energy system of the Drake Landing Solar Community in Okotoks, Alberta, Canada, showing the main system components and the primary flows of energy.

Table 4.9 Comparison of energy use and greenhouse gas (GHG) emissions associated with heating (space and domestic hot water) for conventional homes and those in the Drake Landing Solar Community.

Parameter	Conventional home	Drake Landing Solar Community home	Reduction (relative to conventional home)
Energy use rate (GJ/yr)			
Solar energy	0	71	−71[a]
Natural gas	126	15	111
Total	126	86	40
GHG emission rate (t/yr)	6.4	0.8	5.7

a) The negative reduction in solar energy use reflects the fact that Drake Landing Solar Community homes use solar energy and conventional homes do not.

During operation, a glycol heat transport fluid conveys thermal energy from the solar collectors through insulated pipes to the Energy Centre. There, a heat exchanger transfers heat from the solar collector loop to two thermal storage tanks containing water. Hot water from the tanks can be transferred via a district heating network to DLSC homes. The short-term TESs in the Energy Center act as a buffer between the collector loop, the district energy loop and the BTES field, receiving and discharging thermal energy as necessary. The short-term storage tanks support the system operation by being able to receive and discharge heat at a much greater rate than the BTES storage, which has a much higher TES capacity. Heat is stored in the ground by circulating hot water from the tanks to the BTES in warm months, and water from the boreholes is transferred via the heat exchanger to the district energy network for heating homes in cold months.

The DLSC is described in greater detail as a case study in Section 12.2.2.

4.4 Integration of Thermal Energy Storage with Heat Pumps

The heat pump is an important component of many energy efficiency and conservation strategies. TES can be used beneficially in conjunction with heat pump technology, for heating and cooling applications.

4.4.1 Applications of Heat Pumps with Thermal Energy Storage

Several organizations describe applications of heat pump and thermal storage technologies, as well as their integration. Ground-source heat pumps have gained noteworthy market shares in Sweden, Japan, and other countries with similar climatic conditions (Forsen 2005; Nordell and Ahlstrom 2007; Halozan 2008). Sweden has over 300 000 ground-source heat pumps, providing over 27% of space heating needs. The Canadian GeoExchange Coalition (2007) notes that the Canadian market is small presently, with only 3000 units installed in 2006. For context, it is noted that the Canadian building sector accounts for 31% of total secondary energy use and 28% of GHG emissions (Marbek Resource Consultants Ltd 1999; Caneta 2003, 2004; Hanova and Dowlatabadi 2007). Space heating is responsible for approximately 55% of the energy use and GHG emissions, and air conditioning use is also significant, accounting for peak electrical demand during the summer in many regions in Canada, including Ontario.

Research is needed to increase applications (Marbek Resource Consultants Ltd 1999; Spitler 2005), and several investigations aimed at improving ground-source heat pump technical and economic performance and developing alternative configurations were reported at the International Energy Agency's International Heat Pump Conferences (e.g., Halozan 2008).

Several organizations are actively promoting heat pump and thermal storage technologies. Some significant examples are:

- The IEA Heat Pump Center (www.heatpumpcentre.org) is an international information service center for heat pumping technologies, applications and markets. The goal of the Center is to accelerate the implementation of heat pumps and related heat pumping technologies, including air conditioning and refrigeration. The Center publishes the IEA Heat Pump Center Newsletter, which often has articles on TES (e.g., Van de Ven 1999).

- The CANMET Energy Technology Centre (CETC) (www.nrcan.gc.ca/es/etb/cetc) is a Canadian federal government science and technology organization with a mandate to develop and demonstrate energy efficient, alternative and renewable energy technologies and processes, including heat pumps and TES (CANMET 2005).
- The Canadian GeoExchange Coalition is an industrial association representing hundreds of members that promotes the ground-based heat pump industry actively (Canadian GeoExchange Coalition 2007).

Heat pumps and TES can be integrated in various ways, and such a combination can be beneficial for heating and/or cooling applications. These benefits are discussed in the next few subsections.

4.4.2 Benefits of Integrating Heat Pumps with Thermal Energy Storage for Heating

Heat pumps and TES can be integrated in a variety of ways for heating applications. The benefits of such a combination include the following:

- Heating with heat pumps and TES can be accomplished using ground-based storage, where the heat source to the heat pump is the ground.
- Heat pumps can also be combined with latent TES. For example, a heat pump can use a latent TES to enable rapid room temperature increases and defrosting. In one such system, the latent TES uses polyethylene glycol as a phase change material, which surrounds a rotary compressor of the air conditioner/heat pump for a room. Heat released from the compressor is transferred to the TES through a finned-tube heat exchanger, and recovered for use during start-up and for defrosting. During start-up, the TES halves the time to reach a 45 °C discharge air temperature. The integration improves heat capacity by about 10% and coefficient of performance by 5%, and requires the same installation space as a conventional air conditioner/heat pump.
- Heat from solar collectors, which is often used directly for space heating, can instead be used as a heat source for a heat pump. In such a solar-augmented heat-pump system, the solar-collector outlet temperatures can be lower than with direct heating, increasing energy efficiency and reducing the cost of the solar collector. The higher source temperature also increases the coefficient of performance of the heat pump, reducing its electricity consumption. Such a system can operate in various modes. With small TES, the solar collector improves heat-pump efficiency mainly during sunny periods. With larger TES, the solar energy provides a warm storage for heat-pump operations during cloudy periods and the night. Alternatively, the overall system can be designed so that the heat pump operates only during off-peak hours. Such an approach requires two TESs, one to store solar energy and one to store the heat-pump output for space heating at all times.

4.4.3 Benefits of Integrating Heat Pumps with Thermal Energy Storage for Cooling

Heat pumps can be beneficially integrated with both latent and sensible TES for cooling applications. Both are considered below.

In the former case, heat pumps can be combined with latent TES using ice and ice slurry. Adding ice to water to create an ice slurry enhances the sensible cooling capacity of water with the latent heat capacity of ice particles.

In the latter case, an integrated system of water-based cold TES combined with heat pumps can:

- make heat pump operation more economic;
- level electrical loads.

In conventional air conditioning using heat pumps, the heat pumps operate during the day when cooling demand exists. This operation contributes to electricity daytime demand, which is significant since cooling demand is sometimes responsible for more than one-third of peak electrical demand. In a typical water-based cold TES system, half of the daily cooling load can met by night operation of heat pumps.

Combining a heat pump and a cold TES can provide the following benefits:

- **Efficient operation of heat pumps:** Although a heat pump typically achieves maximum efficiency at a certain operating condition, heat pump operation normally cannot be maintained at the most efficient condition in commercial and residential applications because cooling and heating loads vary temporally. This variation reduces the seasonal efficiency of heat pumps. Because heat pumps integrated with cold TES can operate independently of the cooling and heating load of buildings, cold TES helps avoid this problem by permitting operation of heat pumps at the most efficient operating condition.
- **Load levelling of air conditioning electricity demand:** Cold TES can shift peak air conditioning loads to the night, increasing significantly the annual dependence on night time electricity. A typical cold TES, for instance, shifts half the air conditioning load to the night on the peak day, and permits the annual dependence on night time electricity to reach up to 70% for cooling and 90% for heating. Also, a cold TES system can improve the annual load factor of electricity generation facilities, and save money for consumers when electric power companies provide discount rates for night time electricity.
- **Reduced heat pump size:** For a fixed air conditioning load, the longer operating hours of a heat pump integrated with cold TES allow a smaller capacity heat pump to be used, reducing electrical demand peaks and decreasing initial and operating costs.

4.4.4 Multi-Season Integration of Heat Pumps with Thermal Energy Storage for Heating and Cooling

Combined heating and cooling operations can also be advantageously accomplished by integrating heat pumps with TES, particularly for multi-season applications.

For example, with a ground-source heat pump with seasonal TES, the ground or groundwater is cooled during heating, as heat from the TES is supplied to the building. After the heating season, the stored cold is used for direct cooling, via cold ground-water from the injection well or cold brine from earth heat exchangers.

After some time, the ground temperature may be too high for direct cooling. The system then can be operated as a conventional heat pump, cooling the building space and storing heat in the ground until the next heating season. Also, a heat pump and an aquifer TES allows cold water to be extracted from a cold well during summer and warmed by cooling a building, and then returned to a warm well in the aquifer. A heat pump can cool the cold water further, if necessary. The warmed water increases the temperature of the aquifer near the warm well. The operation is reversed during winter, with warm

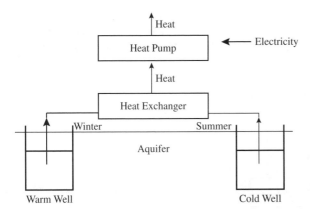

Figure 4.8 An electrically driven ground-source heat pump integrated with an aquifer thermal energy storage, showing both winter and summer operation modes. Heated fluid is extracted from the warm well in winter, while cooled fluid is extracted from the cold well in summer. The heat exchanger and heat pump are shown in winter mode, where they deliver heat for heating. In summer mode, however, the heat pump extracts heat from the space being cooled and rejects the heat via the heat exchanger to the warm well.

water extracted from the warm well and boosted in temperature by the heat pump if necessary.

The combination of a ground-source heat pump with aquifer thermal storage is shown in Figure 4.8. There, two seasonal operating strategies are shown:

- **Summer.** Cold water is extracted from a cold well in the aquifer to cool a building directly or be further cooled first by a heat pump. The heated water is input to a warm well where it raises the aquifer temperature near that well.
- **Winter.** Heat is extracted from the warm well and its temperature is raised if necessary with a heat pump to heat a building.

When heating a building with a ground-source heat pump, the groundwater and/or ground is cooled. Several operation modes of a ground-source heat pump with seasonal thermal storage are possible:

- **Heat pump heating mode (Figure 4.9a).** A heat pump extracts heat from the groundwater or ground and heats a building, cooling the groundwater and/or ground.
- **Direct cooling mode.** At the end of the heating season, enough cold is stored to operate a cooling system directly with cold groundwater from the injection well or cold brine from earth heat exchangers.
- **Heat pump cooling mode (Figure 4.9b).** The heat pump in chiller mode is operated to help meet cooling loads. This operation warms the ground, permitting the system to be operated only as a conventional heat pump plant. Heat is stored in the groundwater or ground for the next heating season.

Although the operating costs of such a system depend on the load characteristics, the most cost-effective and efficient performance is normally achieved by operating the system in heating mode and direct cooling mode.

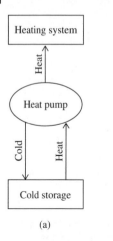
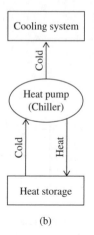

Figure 4.9 Operating strategies for a ground-source heat pump integrated with an aquifer thermal storage. (a) Heating; (b) cooling.

4.4.5 Example: Institutional Heating and Cooling Using Heat Pumps and Thermal Energy Storage

A BTES system is utilized at the north campus of the University of Ontario Institute of Technology, Oshawa, Ontario, Canada, which opened in 2003 and has about 10 000 students (Dincer and Rosen 2011). Most of its buildings are designed to be heated and cooled using ground-source heat pumps in concert with the BTES and its 384 bore-holes, each 213 m deep (Figure 4.10). Thermal energy is upgraded by ground-source heat pumps for building heating, while the ground can receive energy and be warmed using the heat pump in its cooling mode. Fluid circulating through tubing extended into the wells collects heat from the ground and carries it into buildings in winter, while the system extracts heat from buildings and transfers it into the ground in summer. The BTES thereby provides heating and cooling on a seasonal basis. The ground is a useful heat source and heat sink, and energy storage medium, since its temperature is nearly constant throughout the year (except for the upper 5 m).

Chillers pump energy from the buildings into the TES. The heat pump modules assist in this cooling. Chilled water is supplied from two chillers, each having seven 90-t modules, and two sets of heat pumps with seven 50-t modules each. The condensing water passes to the borefield, which stores it for use in the winter (when the heat pumps reverse) and provides low-temperature (53 °C) hot water for the campus. Each building has an internal distribution system and is isolated hydronically with a heat exchanger. Supplemental heating is provided by condensing boilers. In autumn, energy is recovered from the BTES field, and the return water is sufficiently hot for "free-heating" (heating without using the heat pumps). Technical specifications of the chillers and two heat pumps, each having seven modules, are listed in Table 4.10.

The total cooling load of the campus buildings was anticipated to be about 7000 kW. Thermal conductivity test results (Beatty and Thompson 2004) showed the thermal conductivity for the geologic media in a test well to be 1.9 W/m K. This information determined the requirements for the borehole field in terms of numbers and depths of boreholes to meet the energy service needs. Steel casing was installed in the upper 58 m of each borehole to seal out groundwater in the shallow formations. Water-filled borehole heat exchangers (BHEs) were used improve system efficiency

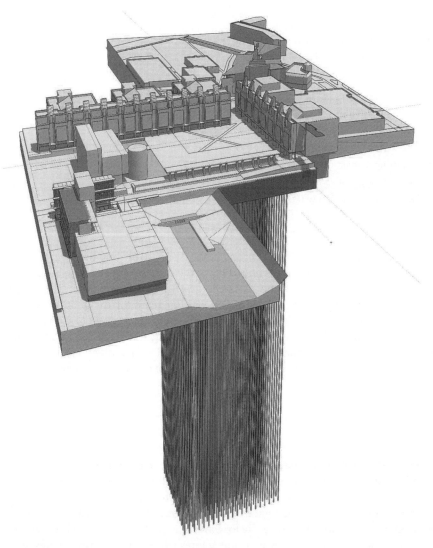

Figure 4.10 Borehole thermal energy storage system at the University of Ontario Institute of Technology, Oshawa, Ontario, Canada. The boreholes comprising the thermal energy storage system are located below the university quadrangle field, which is surrounded by university buildings.

and extend borehole lifetime. The hydrogeologic setting at the vicinity of the site has unconsolidated overburden deposits overlying shale bedrock (Figure 4.6), and the background site temperature is 10 °C. Groundwater resources in the Oshawa area are limited to isolated, thin sand deposits (Beatty and Thompson 2004). The homogeneous, non-fractured rock at the site is suitable for TES since little groundwater flow exists to transport thermal energy from the site. The BTES field has a volume of approximately 1 400 000 m^3, and contains 1700 kt of rock and 600 kt of overburden. The BHEs are located on a 4.5 m grid and the total field is about 7000 m^2 in area.

Table 4.10 Design parameter values for heat pumps integrated the borehole thermal energy storage, broken down by heating and cooling mode, at University of Ontario Institute of Technology, Oshawa, Ontario, Canada.

Parameter	Heating	Cooling
Coefficient of performance	2.8	4.9
Energy load (kW)	1390	1240
Load water entering temperature (°C)	41.3	14.4
Load water exiting temperature (°C)	52.0	5.5
Source water entering temperature (°C)	9.3	29.4
Source water exiting temperature (°C)	5.6	35.0

Numerous underground TES applications exist (IEA 2016), but the University of Ontario Institute of Technology borehole TES is the largest and deepest in Canada, and the geothermal well field is one of the largest in North America. The simple payback period when the system was designed was 7.5 years for the geothermal well field, and 3–5 years for the high-efficiency HVAC equipment. Annual energy costs are reduced using the BTES system by 40% for heating and 16% for cooling, and the system provides indirect financial benefits (reduced boiler plant costs, reduced annual use of potable water and treatment chemicals, avoided costs of roof cooling towers).

4.5 Closing Remarks

Some types of TES technology (sensible and some latent) have reached a high level of maturity, while others (thermochemical) are still in development. Although many types of thermal storage for heat or cold, have been demonstrated and are commercially available, advances are continually being made. Phase change and thermochemical materials have received increased attention recently for various applications, due to their greater volumetric TES capacity at relatively constant temperature. The selection or design of thermal storage for an application is normally based on the application characteristics, and successful utilization normally requires acceptable economics, in terms of capital and operating costs. TES can facilitate the use of intermittent sources of thermal energy such as solar energy and help in addressing problems relating to energy supply, environmental impact, and sustainability. The examples presented of major TES installations illustrate the points covered in this chapter.

References

Agyenim, F., Hewitt, N., Eames, P. and Smyth, M. (2010) A review of materials, heat transfer and phase change problem formulation for latent heat thermal energy storage systems (LHTESS). *Renewable and Sustainable Energy Reviews*, **14**, 615–628.

Air Conditioning Contractors of America (ACCA) (2005) *Thermal Energy Storage: A Guide for Commercial HVAC Contractors*, Air Conditioning Contractors of America, ACCA, Arlington.

Alkan, C. and Sari, A. (2008) Fatty acid/poly (methyl methacrylate) (PMMA) blends as form-stable phase change materials for latent heat thermal energy storage. *Solar Energy*, **82**, 118–124.

Allegrini, J., Orehounig, K., Mavromatidis, G. *et al.* (2015) A review of modelling approaches and tools for the simulation of district-scale energy systems. *Renewable and Sustainable Energy Reviews*, **52**, 1391–1404.

American Society of Heating, Refrigerating and Air-Conditioning Engineers (ASHRAE) (2000) *Standard 943: Method of Testing Active Sensible TES Devices Based on Thermal Performance*, American Society of Heating, Refrigerating and Air-Conditioning Engineers, Atlanta.

American Society of Heating, Refrigerating and Air-Conditioning Engineers (ASHRAE) (2007) *Thermal storage, in ASHRAE Handbook: HVAC Applications*, American Society of Heating, Refrigerating and Air-Conditioning Engineers, Atlanta, Chapter 34.

Arteconi, A., Hewitt, N. and Polonara, F. (2012) State of the art of thermal storage for demand-side management. *Applied Energy*, **93**, 371–389.

Ataer, O.E. (2006) Storage of thermal energy, in *Energy Storage Systems*, vol. **1** (ed. Y.A. Gogus), Encyclopedia of Life Support Systems, EOLSS Publishers, Oxford, pp. 97–116.

Azpiazu, M.N., Morquillas, J.M. and Vazquez, A. (2003) Heat recovery from a thermal energy storage based on the $Ca(OH)_2/CaO$ cycle. *Applied Thermal Engineering*, **23**, 733–741.

Bakker, M., van Helden, W.G.J., and Hauer, A. (2009) *Materials for compact thermal energy storage: A new IEA joint SHC/ECES task. Proceeding of the ASME 2009 3rd International Conference of Energy Sustainability*, July 19–23, 2009, San Francisco, USA. ASME.

Bal, L.M., Satya, S. and Naik, S.N. (2010) Solar dryer with thermal energy storage systems for drying agricultural food products: A review. *Renewable and Sustainable Energy Reviews*, **14**, 2298–2314.

Bales, C. (2006) Chemical and sorption heat storage. Proceedings of the DANVAK Seminar, DANVAK Seminar (Solar Heating Systems – Combisystems - Heat Storage), November 14, 2006, Lyngby, Denmark. TechMedia A/S.

Bayés-García, L., Ventolà, L., Cordobilla, R. *et al.* (2010) Phase change materials (PCM) microcapsules with different shell compositions: Preparation, characterization and thermal stability. *Solar Energy Materials and Solar Cells*, **94**, 1235–1240.

Beatty, B. and Thompson, J. (2004) 75 km of drilling for thermal energy storage. *Geo-Engineering for the Society and its Environment: Proceedings of the 57th Canadian Geotechnical Conference*, Session 8B, October 24–27, 2004, Quebec City, Canada, pp. 38-43.

Beckman, G. and Gilli, P.V. (1984) *Thermal Energy Storage*, Springer-Verlag, New York.

Bruno, F., Belusko, M., and Tay, N.H.S. (2010) Design of PCM thermal storage system using the effectiveness-NTU Method. *Proceedings of the EuroSun 2010 Conference*, September 28–October 1, 2010, Graz, Austria. International Solar Energy Society (ISES), paper 200.

Cabeza, L.F., Castell, A., Barreneche, C. *et al.* (2011) Materials used as PCM in thermal energy storage in buildings: A review. *Renewable Sustainable Energy Review*, **15** (3), 1675–1695.

Cabeza, L.F., Castellón, C., Nogués, M. *et al.* (2007) Use of microencapsulated PCM in concrete walls for energy savings. *Energy and Buildings*, **39**, 113–119.

Canadian GeoExchange Coalition (2007) *Survey of Canadian Geoexchange Industry: 2004-2006*. GeoConneXion Magazine, December, pp. 10–13.

Canadian Standards Association (CSA) (2002) *Design and Installation of Underground Thermal Energy Storage Systems for Commercial and Institutional Buildings, Standard CAN/CSA-C448-02*, Canadian Standards Association CSA, Mississauga.

Caneta (2003) *Global Warming Impacts of Ground-Source Heat Pumps Compared to Other Heating and Cooling Systems*. Final Report. Caneta Research Inc., Mississauga.

Caneta (2004) *Market, Economic, and Barrier Analysis for Ground Source Heat Pumps in Canada, US, and Europe*. Draft Report. Caneta Research Inc., Mississauga.

CANMET (2005) *Ground-source heat pump project analysis, in Clean Energy Project Analysis*, CANMET Energy Technology Centre – Varennes, Minister of Natural Resource Canada, Ottawa.

Cui, P., Diao, N., Gao, C. and Fang, Z. (2015) Thermal investigation of in-series vertical ground heat exchangers for industrial waste heat storage. *Geothermics*, **57**, 205–212.

Desgrosseilliers, L., Whitman, C.A., Groulx, D. and White, M.A. (2013) Dodecanoic acid as a promising phase-change material for thermal energy storage. *Applied Thermal Engineering*, **53**, 37–41.

Dickinson, R.M., Cruickshank, C.A. and Harrison, S.J. (2013) Charge and discharge strategies for a multi-tank thermal energy storage. *Applied Energy*, **109**, 366–373.

Dincer, I. and Rosen, M.A. (2001) Energetic, environmental and economic aspects of thermal energy storage systems for cooling capacity. *Applied Thermal Engineering*, **21**, 1105–1117.

Dincer, I. and Rosen, M.A. (2007) *Exergy: Energy, Environment, and Sustainable Development*, Elsevier, Oxford.

Dincer, I. and Rosen, M.A. (2011) *Thermal Energy Storage: Systems and Applications*, 2nd edn, John Wiley & Sons, Ltd, Chichester.

Dinter, F., Ger, M. and Tamme, R. (1991) *Thermal Energy Storage for Commercial Applications*, Springer-Verlag, Berlin.

Dominkovic, D.F., Cosic, B., Bacelic Medic, Z. and Duic, N. (2015) A hybrid optimization model of biomass trigeneration system combined with pit thermal energy storage. *Energy Conversion and Management*, **104**, 90–99.

Eslami-nejad, P. and Bernier, M. (2013) A preliminary assessment on the use of phase change materials around geothermal boreholes. *ASHRAE Tranactions*, **119** (2), 312–321.

Fan, L. and Khodadadi, J.M. (2011) Thermal conductivity enhancement of phase change materials for thermal energy storage: A review. *Renewable and Sustainable Energy Reviews*, **15**, 24–46.

Fang, G., Li, H., Liu, X. and Wu, S. (2010) Experimental investigation of performances of microcapsule phase change material for thermal energy storage. *Chemical Engineering & Technology*, **33**, 227–230.

Farid, M. and Kong, W.J. (2001) Underfloor heating with latent heat storage. *Proceedings of the Institution of Mechanical Engineers, Part A: Journal of Power and Energy*, **215**, 601–609.

Farid, M.M., Khudhair, A.M., Razack, S.A.K. and Al-Hallaj, S. (2004) A review on phase change energy storage: Materials and applications. *Energy Conversion and Management*, **4**, 1597–1615.

Fernandez, A.I., Martínez, M., Segarra, M. *et al.* (2010) Selection of materials with potential in sensible thermal energy storage. *Solar Energy Materials and Solar Cells*, **94**, 1723–1729.

Flynn, C. and Siren, K. (2015) Influence of location and design on the performance of a solar district heating system equipped with borehole seasonal storage. *Renewable Energy*, **81**, 377–388.

Forsen, M. (2005) *Heat Pumps – Technology and Environmental Impact (Part 1)*. Report. Swedish Heat Pump Association, Stockholm.

Giordano, N., Comina, C., Mandrone, G. and Cagni, A. (2016) Borehole thermal energy storage (BTES). First results from the injection phase of a living lab in Torino (NW Italy). *Renewable Energy*, **86**, 993–1008.

Guadalfajara, M., Miguel, A., Lozano, L. and Serra, L. (2014) Comparison of simple methods for the design of central solar heating plants with seasonal storage. *Energy Procedia*, **48**, 11107–11115.

Haji Abedin, A. and Rosen, M.A. (2010a) Energy and exergy analyses of a closed thermochemical energy storage system: Methodology and illustrative application. *Proceedings of the. 23rd Conference on Efficiency, Cost, Optimization, and Environmental Impact of Energy Systems*, June 14–17, 2010, Lausanne, Switzerland, Vol. **III**, pp. 107–114.

Haji Abedin, A. and Rosen, M.A. (2010b) Energy and exergy analysis of a closed thermochemical energy storage system. *Proceedings of the 5th International Green Energy Conference (IGEC-V)*, June 1–3, 2010, Waterloo, Canada, pp. 1-17.

Haller, M.Y., Cruickshank, C.A., Streicher, W. *et al.* (2009) Methods to determine stratification efficiency of thermal energy storage processes: Review and theoretical comparison. *Solar Energy*, **83**, 1847–1860.

Halozan, H. (2008) Ground-source heat pumps and buildings. *Proceedings of the 9th International IEA Heat Pump Conference*, May 20–22, 2008, Zurich, Switzerland. International Energy Agency.

Hamada, Y., Ohtsu, W. and Fukai, J. (2003) Thermal response in thermal energy storage material around heat transfer tubes: effect of additives on heat transfer rates. *Solar Energy*, **75**, 317–328.

Hanova, J. and Dowlatabadi, H. (2007) Strategic GHG reduction through the use of ground source heat pump technology. *Environmental Research Letters*, **2** (**4**), 1–8.

Hasnain, S.M. (1998) Review of thermal energy storage. Part I: Heat storage materials and techniques. *Energy Conversion and Management*, **39**, 1127–1138.

Hasnain, S.M., Smiai, M., Al-Saedi, Y. and Al-Khaldi, M. (1996) *Energy Research Institute – Internal Report*, KACST, Riyadh.

Hauer, A. and Lavemann, E. (2007) *Open absorption systems for air conditioning and thermal energy consumption, Part VI*, in *Thermal Energy Storage for Sustainable Energy Consumption* (ed. H.O. Paskoy), Springer, Dordrecht, pp. 429–444.

Heier, J., Bales, C. and Martin, V. (2015) Combining thermal energy storage with buildings – a review. *Renewable and Sustainable Energy Reviews*, **42**, 1305–1325.

Heim, D. (2010) Isothermal storage of solar energy in building construction. *Renewable Energy*, **35**, 788–796.

Heinz, A. and Schranzhofer, H. (2010) *Thermal energy storage with phase change materials: A promising solution? Proceedings of the EuroSun 2010 Conference*, September 28–October 1, 2010, Graz, Austria. International Solar Energy Society (ISES), paper 206.

Henze, G.P., Krarti, M. and Brandemuehl, M.J. (2003) Guidelines for improved performance of ice storage systems. *Energy and Buildings*, **35**, 111–127.

Hill, J.E., Kelly, G.E. and Peavy, B.A. (1977) A method of testing for rating thermal storage devices based on thermal performance. *Solar Energy*, **19**, 721–732.

Hongois, S., Kuznik, F., Stevens, P., Roux, J.J., Radulescu, M., and Beuarepaire, E. (2010), Thermochemical storage using composite materials: From the material to the system. *Proceedings of the Eurosun 2010 Conference*, September 28–October 1, 2010, Graz, Austria. International Solar Energy Society (ISES).

International Energy Agency (IEA) (2008) Compact Thermal Energy Storage: Material Development and System Integration, in *Technical Report (draft), Task 42, Annex 28, Solar Heating and Cooling Programme*, International Energy Agency, Paris.

International Energy Agency (IEA) (2009) Energy Conservation through Energy Storage. http://www.iea-eces.org (accessed August 14, 2009).

International Energy Agency (IEA) (2010) *Energy Technology Perspectives 2010: Scenarios and Strategies to 2050*, Report, International Energy Agency, Paris.

International Energy Agency (IEA) (2016) Energy Conservation through Energy Storage. http://www.iea-eces.org (accessed March 16, 2016).

Jegadheeswaran, S. and Pohekar, S.D. (2009) Performance enhancement in latent heat thermal storage system: A review. *Renewable and Sustainable Energy Reviews*, **13**, 2225–2244.

Khalifa, A.J.N. and Abbas, E.F. (2009) A comparative performance study of some thermal storage materials used for solar space heating. *Energy and Buildings*, **41**, 407–415.

Koca, A., Oztop, H., Koyun, T. and Varol, Y. (2008) Energy and exergy analysis of a latent heat storage system with phase change material for a solar collector. *Renewable Energy*, **33**, 567–574.

Koohi-Fayegh, S. and Rosen, M.A. (2014) An analytical approach to evaluating the effect of thermal interaction of geothermal heat exchangers on ground heat pump efficiency. *Energy Conversion and Management*, **78**, 184–192.

Kreetz, H. and Lovergrove, K. (1999) Theoretical analysis and experimental results of a 1 kW_{chem} ammonia synthesis reactor for a solar thermochemical energy storage system. *Solar Energy*, **67**, 287–296.

Lahmidi, H., Mauran, S. and Goetz, V. (2006) Definition, test and simulation of a thermochemical storage process adapted to solar thermal systems. *Solar Energy*, **80**, 883–893.

Lai, C., Chen, R.H. and Lin, C. (2010) Heat transfer and thermal storage behaviour of gypsum boards incorporating micro-encapsulated PCM. *Energy and Buildings*, **42**, 1259–1266.

Lee, K.S. (2010) A review on concepts, applications, and models of aquifer thermal energy storage systems. *Energies*, **3** (6), 1320–1334.

Lovergrove, K., Luzzi, A. and Kreetz, H. (1999a) A solar-driven ammonia-based thermochemical energy storage system. *Solar Energy*, **67**, 309–316.

Lovergrove, K., Luzzi, A., McCann, M. and Freitag, O. (1999b) Exergy analysis of ammonia-based solar thermochemical power systems. *Solar Energy*, **66**, 103–115.

Marbek Resource Consultants Ltd (1999) Ground Source Heat Pump Market Development Strategy.

Masruroh, N.A., Li, B. and Klemes, J. (2006) Life cycle analysis of a solar thermal system with thermochemical storage process. *Renewable Energy*, **31**, 537–548.

Mathur, A., Kasetty, R., and Hardin, C. (2010) A practical phase change thermal energy storage for concentrating solar power plants. *Proceedings of the EuroSun 2010 Conference*, September 28–October 1, 2010, Graz, Austria. International Solar Energy Society (ISES), paper 217.

Mauran, S., Lahmidi, H. and Goetz, V. (2008) Solar heating and cooling by a thermochemical process: First experiments of a prototype storing 60 kWh by a solid/gas reaction. *Solar Energy*, **82**, 623–636.

McClenahan, D., Gusdorf, J., Kokko, J. *et al.* (2006) *Okotoks: Seasonal Storage of Solar Energy for Space Heat in a New Community*, ACEEE, 2006 Summer Study on Energy Efficiency in Buildings, Pacific Grove, CA.

Nordell, B. and Ahlstrom, A.-M. (2007) *Freezing problems in borehole heat exchangers, in Thermal Energy Storage for Sustainable Energy Consumption: Proceedings of the NATO Advanced Study Institute on Thermal Energy Storage for Sustainable Energy Consumption – Fundamentals, Case Studies and Design* (ed. H.O. Paksoy), NATO Science Series II: Mathematics, Physics and Chemistry, Springer, Dordrecht, pp. 193–204.

Novo, A.V., Bayon, J.R., Castro-Fresno, D. and Rodriguez-Hernandez, J. (2010) Review of seasonal heat storage in large basins: Water tanks and gravel–water pits. *Applied Energy*, **87**, 390–397.

Paksoy, H.O. (ed.) (2007) *Thermal Energy Storage for Sustainable Energy Consumption: Fundamentals, Case Studies and Design, NATO Science Series II: Mathematics, Physics and Chemistry*, vol. **234**, Springer, New York.

Pavlov, G.K. and Olesen, B.W. (2012) Thermal energy storage - A review of concepts and systems for heating and cooling applications in buildings: Part 1 - Seasonal storage in the ground. *HVAC&R Research*, **18** (**3**), 515–538.

Piette, M.A. (1990) *Learning from Experience with Diurnal Thermal Energy Storage Managing Electric Loads in Buildings, Analysis Support Unit, Centre for Analysis and Dissemination of Demonstrated Energy Technologies (CADDET)*, International Energy Agency, Sittard.

Pinel, P., Cruickshank, C.A., Beausoleil-Morrison, I. and Wills, A. (2011) A review of available methods for seasonal storage of solar thermal energy in residential applications. *Renewable & Sustainable Energy Reviews*, **15**, 3341–3359.

Rad, F.M., Fung, A.S. and Leong, W.H. (2013) Feasibility of combined solar thermal and ground source heat pump systems in cold climate, Canada. *Energy and Buildings*, **61**, 224–232.

Réveillèrea, A., Hamma, V., Lesueur, H. *et al.* (2013) Geothermal contribution to the energy mix of a heating network when using Aquifer Thermal Energy Storage: Modeling and application to the Paris basin. *Geothermics*, **47**, 69–79.

Rezaie, B., Reddy, B.V. and Rosen, M.A. (2015) Exergy analysis of thermal energy storage in a district energy application. *Renewable Energy*, **74**, 848–854.

Rosen, M.A. (2015) *Thermal storage, in Alternative Energy and Shale Gas Encyclopedia* (eds J.H. Lehr and J. Keeley), John Wiley & Sons, Inc., Hoboken, Section 3, Chapter 8.

Rosen, M.A., Tang, R. and Dincer, I. (2004) 'Effect of stratification on energy and exergy capacities in thermal storage systems. *Inernational Journal of Energy Research*, **28**, 177–193.

Science Applications International Corporation (SAIC) (2010) *Drake Landing Solar Community Energy: Annual Report for 2008–2009*, Report CM002171, Science Applications International Corporation, Ottawa.

Saito, A. (2002) Recent advances in research on cold thermal energy storage. *International Journal of Refrigeration*, **25**, 177–189.

Sanner, B. and Chant, V.G. (1992) Seasonal cold storage in the ground using heat pumps. *IEA Heat Pump Center Newsletter*, **10**(1), 4–7.

Sanner, B., Karytsas, C., Mendrinos, D. and Rybach, L. (2003) Current status of ground source heat pumps and underground thermal energy storage in Europe. *Geothermics*, **32** (**4–6**), 579–588.

Sari, A. and Karaipekli, A. (2009) Preparation, thermal properties and thermal reliability of palmitic acid/expanded graphite composite as form-stable PCM for thermal energy storage. *Solar Energy Materials and Solar Cells*, **93**, 571–576.

Scalat, S., Banu, D., Hawes, D. *et al.* (1996) Full scale thermal testing of latent heat storage in wallboard. *Solar Energy Materials and Solar Cells*, **44**, 49–61.

Swiss Federal Energy Office (SFEO) (1990) *Guide to Seasonal Storage*, Swiss Federal Energy Office, Ittigen.

Sharma, A., Tyagi, V.V., Chen, C.R. and Buddhi, D. (2009) Review on thermal energy storage with phase change materials and applications. *Renewable and Sustainable Energy Reviews*, **13**, 318–345.

Shilei, L., Guohui, F., Neng, Z. and Li, D. (2007) Experimental study and evaluation of latent heat storage in phase change materials wallboards. *Energy and Buildings*, **39**, 1088–1091.

Sibbitt, B., McClenahan, D., Djebbar, R. *et al.* (2012) The performance of a high solar fraction seasonal storage district heating system – five years of operation. *Energy Procedia*, **30**, 856–865.

Sibbitt, B., Onno, T., McClenahan, D., Thornton, J., Brunger, A., Kokko, J., and Wong, B. (2007) *The Drake Landing Solar Community project: Early results. Proceedings of the Canadian Solar Buildings Conference*, June 10–14, 2007, Calgary, Canada.

Singh, H., Saini, R.P. and Saini, J.S. (2010) A review on packed bed solar energy storage systems. *Renewable and Sustainable Energy Reviews*, **14**, 1059–1069.

Smart Net-zero Energy Buildings Strategic Research Network (SNEBRN) (2015) http://www.solarbuildings.ca/index.php/en/home/about-us (accessed July 26, 2016).

Spitler, J.D. (2005) Ground-source heat pump system research – past, present, and future. *International Journal of HVAC&R*, **11** (2), 165–167.

Tatsidjodoung, P. and Le Pierrès, N. (2013) A review of potential materials for thermal energy storage in building applications. *Renewable and Sustainable Energy Reviews*, **18**, 327–349.

Terziotti, L.T., Sweet, M.L. and McLeskey, J.T. Jr., (2012) Modeling seasonal solar thermal energy storage in a large urban residential building using TRNSYS 16. *Energy and Buildings*, **45**, 28–31.

Van de Ven, H. (1999) Status and trends of the European heat pump market. *IEA Heat Pump Center Newsletter*, **17** (1), 10–12.

van Essen, V.M., Bleijendaal, L.P.J., Kikkert, B.W.J., Zondag, H.A., Bakker, M., and Bach, P.W. (2010) *Development of a compact heat storage system based on salt hydrates. Proceedings of the Eurosun 2010 Conference*, September 28–October 1, 2010, Graz, Austria. International Solar Energy Society (ISES).

van Helden, W.G.J. (2009) Compact Thermal Energy Storage. http://www.leonardo-energy .org/webinar/compact-thermal-energy-storage-technologies-status-applications-and-developments (accessed July 26, 2016).

Velraj, R., Seeniraj, R.V., Hafner, B. *et al.* (1999) Heat transfer enhancement in a latent heat storage system. *Solar Energy*, **65**, 171–180.

Visscher, K. and Veldhuid, J.B.J. (2005) Comparison of candidate material for seasonal storage of solar heat through dynamic simulation of building and renewable energy system. *Proceedings of the Ninth International IBPSA Conference*, August 15–18, 2005, Montreal, Canada. International Building Performance Simulation Association, pp. 1285-1292.

Weber, R. and Dorer, V. (2008) Long-term heat storage with NaOH. *Vacuum*, **82**, 708–716.

Wettermark, G. (1989) Thermochemical energy storage. *Proceedings of the NATO Advanced Study Institute on Energy Storage Systems*, **167**, 673–681.

Wong, W.P., McClung, J.L., Snijders, A.L., Kokko, J.P., McClenahan, D., and Thornton, J. (2006) First large-scale solar seasonal borehole thermal energy storage in Canada. *Proceedings of Ecostock 2006 Conference*, May 31–June 2, 2006, Stockton, USA. Pomona.

Yan, C., Shi, W., Li, X. and Wang, S. (2016) A seasonal cold storage system based on separate type heat pipe for sustainable building cooling. *Renewable Energy*, **85**, 880–889.

Zalba, B., Marin, J.M., Cabeza, L.F. and Mehling, H. (2004) Free-cooling of buildings with phase change materials. *International Journal of Refrigeration*, **27**, 839–849.

Zondag, H., van Essen, V.M., He, Z., Schuitema, R., and van Helden, W.G.J. (2007) Characterization of $MgSO_4$ for thermochemical storage. *Proceedings of the Second International Renewable Energy Storage Conference (IRES II)*, November 19–21, 2007, Bonn, Germany. Bonn Eurosolar.

Zondag, H.A., Kalbasenka, A., van Essen, M., Bleijendaal, L., van Helden, W., and Krosse, L. (2008) First studies in reactor concepts for thermochemical storage. *Proceedings of the Eurosun 2008 Conference*, October 7–10, 2008, Lisbon, Portugal. International Solar Energy Society (ISES).

5

Geothermal Heating and Cooling

Ground-based energy can provide heating in winter and cooling in summer, in partial or full manners, as explained below:

- Ground-based energy is capable of providing heating in winter, since it is warmer underground than in the air above the ground surface. Sometimes the ground temperature is adequate for heating, but more often it is adequate only for preheating. The temperature of heat extracted from the ground can also be raised using heat pumps, enhancing its ability to be applied for useful heating.
- Ground-based energy can provide direct cooling in summer, since it is cooler underground than in the air above the ground surface. Sometimes the ground temperature is sufficient for cooling, but more often it is only capable of precooling. The temperature of cooling provided by the ground can also be lowered using heat pumps in a cooling mode, enhancing its ability to be applied for useful cooling.

A geothermal heating and cooling system consists of three main components: a heat pump; an underground heat exchanger; and a distribution system such as air ducts. The cost of the system is roughly proportional to the heat exchanger size and, therefore, there is an incentive to evaluate peak monthly and daily loads and average annual loads and to design the heat exchanger as small as reasonably possible to meet the required heat transfer for system operation. Heat exchanger performance is influenced by several factors: the structural and geometric configuration of the heat exchanger; the ground temperature distribution; soil moisture content and its thermal properties; groundwater movement and possible freezing in soil. Thus, appropriate and validated tools are needed, with which the thermal behavior of ground heat exchangers (GHEs) can be assessed and optimized, considering technical, environmental and economic aspects.

In this chapter background information is provided on geothermal systems that provide heating and cooling using the ground. The various uses of the ground as a source of energy or as a storage medium are discussed. General description of ground-source heat pumps (GSHPs), geothermal heat exchangers, their combination with renewable energy sources such as solar sources, and their costs, are also provided.

Geothermal Energy: Sustainable Heating and Cooling Using the Ground, First Edition.
Marc A. Rosen and Seama Koohi-Fayegh.

5.1   Ground-Source Heat Pumps

Parameters affecting the operation of ground-source heat pumps and their efficiencies are discussed in this section. Also covered are ways in which the use of ground-source heat pumps can reduce fossil fuel consumption for heating and cooling applications.

In general, heat pumps operate between a high-temperature medium (heat sink at T_H) and a low-temperature medium (heat source at T_L) (Figure 2.4). A ground-source heat pump, also called a ground-coupled heat pump, geothermal heat pump, or ground heat pump, is a heat pump that uses the ground as its heat source or sink. The heat exchange between the ground and the heat pump cycle occurs via a GHE. If heating or cooling a space, the heat exchange between the heat pump cycle and the space is made via a heating or cooling coil. When cooling a space (ground heat delivery), the temperature of the fluid that runs in the GHE can be considered the high-temperature medium and the cooling coil temperature can be considered the low-temperature medium in the cooling season. In the heating season (ground heat removal), the heating coil temperature can be considered to be the high-temperature medium while the temperature of the fluid running through the heat exchanger can be considered to be the low-temperature medium. In an ideal GSHP where all the processes are reversible, the coefficient of performance of the heat pump can be written as:

$$COP_{cooling} = \frac{1}{1 - \frac{T_{cooling\ coil}}{T_f}} \tag{5.1}$$

$$COP_{heating} = \frac{1}{1 - \frac{T_f}{T_{heating\ coil}}} \tag{5.2}$$

where T_{coil} is the coil temperature, assumed to be constant and within a standard range here. Note that the coefficient of performance in Equation (5.1) and Equation (5.2) can only be used for a reversible heat pump.

Compared with a conventional air-source heat pump, which circulates outdoor air to exchange heat, a GSHP exchanges heat by circulating a fluid in the ground. The ground has a lower temperature than the outdoor air in the cooling mode and a higher temperature than the outdoor air in the heating mode. Consequently, the temperature lift across a GSHP is lower than that of an air-source heat pump for both heating and cooling. Thus, the efficiency of the heat pump, which depends directly on the temperature difference between the borehole fluid and the room, is enhanced for a GSHP. Therefore, due to concern about greenhouse gas emissions and high energy prices, the placement of heat loops in the ground is an increasingly common practice for heating and cooling residential, commercial, institutional, recreational and industrial structures. Low-temperature geothermal energy has the potential to contribute significantly to mitigating both of these problems. Figure 5.1 shows the growth in the GSHP installations from 1996 to 2008 in Canada (Canadian GeoExchange Coalition 2010).

A more detailed comparison between ground- and air-source heat pumps is given in Table 5.1.

In the lower to middle efficiency range, GSHPs use single-speed rotary or reciprocating compressors, relatively standard refrigerant-to-air ratios, but oversized enhanced-surface refrigerant-to-water heat exchangers. Mid-range efficiency heat pump units employ scroll compressors or advanced reciprocating compressors. Units

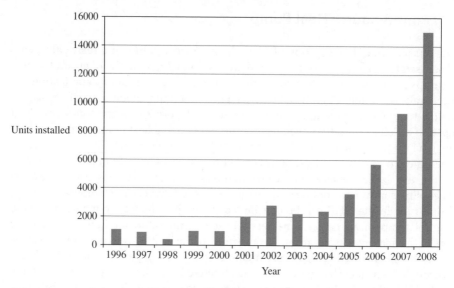

Figure 5.1 Ground-source heat pump installation growth from 1996 to 2008 in Canada.

in the high efficiency range tend to use two-speed compressors or variable speed indoor fan motors or both, with more or less the same heat exchangers (Omer 2008).

The temperature of the ground source and the heating systems used in buildings both affect the type of the heat pump that can meet the space loads. Single-stage heat pumps can provide heat at temperatures of up to 55 °C and are often used to heat buildings that are equipped with low-temperature heating systems such as radiant surfaces. Buildings that use high-temperature heating systems, such as radiators, require heat carriers with temperatures higher than 60 °C. In such cases, two-stage heat pumps with flash vessels are used. The selection of the appropriate refrigerant which could work within the temperature of the geothermal source and the condenser of the second stage where the heat is transferred to the heating system at high temperature could prove challenging. To avoid such challenges, different refrigerants can be used in two single-stage heat pumps that are connected via a heat exchanger (Kulcar *et al.* 2008). The heat exchanger between the two stages operates as a condenser for the first heat pump and as an evaporator for the second (Figure 5.2).

5.2   Geothermal Heat Exchangers

Geothermal heat exchangers, also called GHEs, underground heat exchangers, and ground-coupled heat exchangers, are heat exchangers that exchange heat between a fluid and the ground. Other names exist for such heat exchangers based on more detailed specifications and design variations, although their operation remains fundamentally similar. They are often categorized based on the range of temperature in which they work, that is, low-temperature and high-temperature geothermal heat exchangers. The various types of GHEs are described in this section. Information on systems that use the ground as an energy source or as a storage medium is provided. Depending on

Table 5.1 Qualitative comparison of ground- and air-source heat pumps, broken down by characteristic

Characteristic	Air-source heat pumps	Ground-source heat pumps	
		Vertical	Horizontal
Efficiency	1[a]	3	2
Design criteria			
Feasibility	2	1	1
Construction difficulty	2	3	2
Life cycle cost			
Installation	1	3	2
Operation	2	1	1
Maintenance	2	1	1
Total	2	1	1
Environmental			
CO_2 emissions	2	1	1
Land disturbance	0	2	1
Water contamination	0	0	0
Durability	1	2	2
Practical issues			
Operating restrictions	2	1	1
Aesthetics	1	2	2
Quietness	1	3	3
Vandalism	1	0	0
Safety	2	2	2

a) Numbers represent the importance or significance of the characteristic: 3, high significance; 2, moderate significance; 1, minor significance; 0, no significance.
Source: Adapted from Atam and Helsen (2016).

the type of GHE, they can be coupled to a GSHP or used directly in heating and cooling applications.

5.2.1 Low-Temperature Geothermal Heat Exchangers

Low-temperature geothermal systems also known as geoexchange systems interact closely with the shallow subsurface and have a near-environment temperature. Thermal energy from the ground derives from the background temperature of the ground, which predominantly is dependent on the ambient conditions. Their applications include space heating and cooling systems, aquaculture, agricultural drying, snow melting (particularly in Iceland), and sometimes low wattage power generation, although the overall efficiency of such systems may still not compete with high-temperature geothermal systems used for electrical power generation. Heat exchange between a space and a low-temperature ground source may be done directly or through a heat pump system. In the latter case, the geothermal heat exchanger is built into the heat pump system. GHEs are also classified as open loop (groundwater systems) or closed

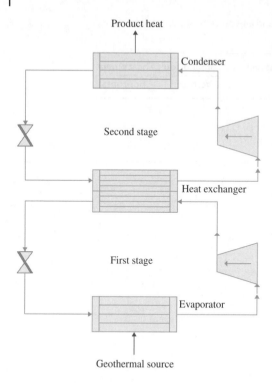

Product heat

Condenser

Second stage

Heat exchanger

First stage

Evaporator

Geothermal source

Figure 5.2 Schematic of a two-stage heat pump and the heat exchanger providing an interface between the stages.

loop (sometimes called ground-coupled systems), with a third category for those not belonging to either. In the following paragraphs, direct-exchange, open-loop and closed-loop GHEs that use the ground as a heat source or sink in a heat pump cycle are briefly explained.

5.2.1.1 Direct Exchange

A direct-exchange loop is a GHE that is in direct contact with the GSHP by circulating the refrigerant in the ground. The heat is exchanged directly between the refrigerant and the ground through copper tubes. Such heat exchange is more efficient than using an additional heat exchanger connected to the condenser or evaporator of the heat pump system. Furthermore, the use of refrigerant allows higher heat transfer rates through phase change of the refrigerant. Therefore, the direct-exchange ground loops are often smaller in size and can incur lower capital costs for a given heat injection or extraction load.

5.2.1.2 Open Loop

Groundwater is a good source or sink for a heat pump cycle due to its constant temperature. In an open-loop system, groundwater from a water-bearing layer is pumped from an aquifer through one well, passed through the heat pump where heat is added to or extracted from a heat carrier and then discharged either onto the surface or to another well in the aquifer. Because the system water supply and discharge are not connected, the loop is "open" (Geothermal Heat Pump Consortium 2013). Standing column wells, mine water or tunnel water are examples for this category. In standing column wells, the majority of the water returns to the source well, eliminating the need for reinjection to another well or surface discharge. In a similar way, open-loop systems can be installed to preheat or precool ambient air flowing through tubes buried in the ground. The air is then heated or cooled by a conventional air conditioning unit before entering the building. In open-loop systems the heat exchanger between the refrigerant and the groundwater is subject to fouling, corrosion, and blockage due to the groundwater chemistry, for example, high iron content (Omer 2008). Furthermore, large flow rates, typically between 9.5×10^{-5} m^3/s and 18.9×10^{-5} m^3/s (1.5–3.0 gal/min) per system cooling ton (Omer 2008), are often required for such systems to operate. Therefore, groundwater quality and availability are important factors to consider when installing such systems.

5.2.1.3 Closed Loop

A closed-loop system uses continuous underground pipe loops placed horizontally or vertically in the ground with both ends of the pipe system connected to the heat pump.

In a horizontal ground heat exchanger (HGHE), a number of plastic pipes are connected either in series or in parallel in a horizontal trench (Figure 5.3). The numbers

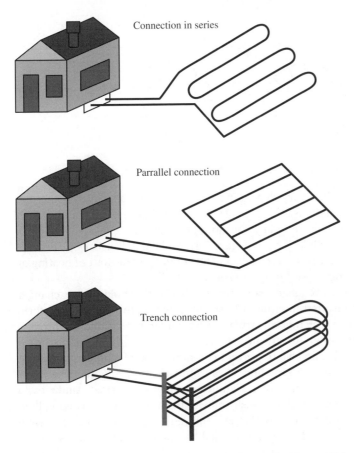

Figure 5.3 Horizontal ground heat exchangers. *Source*: Florides and Kalogirou (2007).

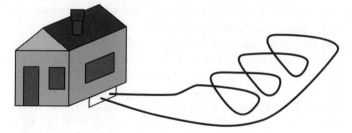

Figure 5.4 Slinky ground heat exchanger. *Source*: Florides and Kalogirou (2007).

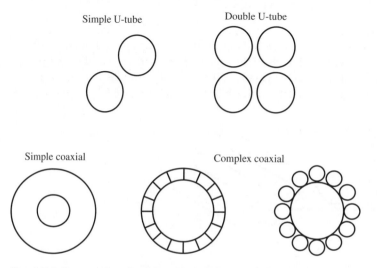

Figure 5.5 Cross section of various types of vertical borehole heat exchangers.

of pipes and trenches installed vary depending on the system capacity and thermal properties of geological formations. This type of GHE is normally most economic when adequate yard space is available. To reduce the amount of land space needed, the pipes are sometimes curled into a "slinky" shape (Figure 5.4). An HGHE is usually placed at a depth of 1–2 m in the ground and is typically 35–60 m long per kilowatt of heating or cooling capacity (Geothermal Heat Pump Consortium 2013).

In a vertical ground heat exchanger, sometimes called a borehole heat exchanger (BHE), plastic pipes are inserted in either a U-shape (called U-tube) or coaxial form (Figure 5.5) into a borehole which is constructed vertically in the ground (Figure 5.6) and is usually filled with grout to enhance thermal contact with the undisturbed ground outside the borehole and also to prevent contamination of aquifers. The grout is often a mixture of sodium bentonite and silica sand which may contain thermally enhanced additives in order to present a significantly higher thermal conductivity than the surrounding ground to facilitate heat transfer from the heat exchanging fluid to the ground and to protect groundwater as required by relevant environmental regulations. The heat carrier fluid, sometimes called borehole fluid, is usually water or water mixed with an environmentally benign antifreeze and flows down to the bottom

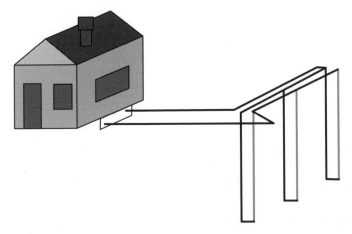

Figure 5.6 Vertical ground heat exchanger. *Source*: Florides and Kalogirou (2007).

of the borehole along one pipe and back upward in another pipe. A typical BHE is usually 20–300 m deep with a diameter of 10–15 cm (Florides and Kalogirou 2007). For high heating or cooling loads, a borehole system composed of a large number of individual boreholes can be installed. The overall system of several boreholes is often called a borefield. The number of boreholes needed and their depth depend mostly on the size of the building, system demands, and the ground temperature. Compared with horizontal heat exchangers, vertical loops are more expensive to install. However, for a given heating and cooling load, they require less piping as the deep ground temperature remains cooler in the summer and warmer in the winter than near-surface ground.

The size of the GHE and the number of pipes depends on the heat exchanger geometry, the capacity of the GSHP coupled to it, the ground temperature, and the properties of the ground.

The distributions of different types of GHEs based on number of installations for some Canadian provinces are shown in Table 5.2. Given the geology and geography, it is seen that the installation of vertical and horizontal ground loop systems varies in different provinces. Overall, closed horizontal loops dominate residential installations in Canada. These systems accounted for 52.5% of residential installations in 2009 while the second largest segment is closed, vertical loops with 34.1% of the installations in the same year (Canadian GeoExchange Coalition 2010).

5.2.2 High-Temperature Geothermal Systems

Other geothermal systems that use the ground as a heat source are described in this section. An example is the direct-exchange geothermal system.

As noted earlier, geothermal energy systems that exploit hot reservoirs in the ground are used for two main purposes:

- electricity generation
- heating

Table 5.2 Distribution of different heat exchanger types based on number of installations in some Canadian provinces.

Province	Open loop (% of provincial systems)	Closed loop (% of provincial systems)		Pond/lake (% of provincial systems)
		Vertical	Horizontal	
Ontario	12	15	67	6
Quebec	6	85	8	1
British Columbia	15	31	52	2
Alberta	7	72	19	2

Source: Adapted from Canadian GeoExchange Coalition (2010).

Some examples of such geothermal sources are:

- ground heated by hot magma
- thermal springs
- geysers

When geothermal energy is extracted, the ground temperature is returned to its elevated temperature by heat contained within hot regions in the earth, via heat and/or mass transfer. This heat originates from the continuing radioactive decay and residual primordial heat from the earth's formation billions of years ago. Such energy is usually deemed to constitute a sustainable resource. This may not be the case when energy is extracted so rapidly that the reservoir does not have time to return to its initial elevated temperature.

Thermal energy extraction from high-temperature underground resources is normally carried out by drilling into the ground.

High-temperature geothermal resources are sometimes categorized by the technology used to extract heat:

- **Hydro-geothermal systems.** In such systems, naturally occurring reservoirs of hot water or steam are utilized that are typically found in permeable rock at less than 5 km below the earth's surface.
- **Enhanced geothermal systems.** In these geothermal systems, water is injected into the non-permeable rocks containing fractures through a well. After being heated, it is forced out from another well (i.e., the production well).

To enhance the performance of the energy extraction method for enhanced geothermal systems, a reservoir can be created or enhanced by injecting pressurized water in the rocks to increase the number and size of the cracks in the rocks. This approach is considered by many to be a technique that will permit significant future enhancements of geothermal energy systems.

High-temperature ground reservoirs can also be categorized by temperature:

- **50–150 °C reservoirs.** Ground reservoirs at 50–150 °C can provide heat for residential and industrial purposes directly, via direct-use systems.
- **>150 °C reservoirs.** Higher temperature ground reservoirs can provide direct heating and electricity generation. The latter is often accomplished with binary, flash steam and dry steam power cycles.

This temperature-based categorization is important since temperature directly correlates with the quality of the energy extracted from the ground.

References

Atam, E. and Helsen, L. (2016) Ground-coupled heat pumps: Part 2. Literature review and research challenges in optimal design. *Renewable and Sustainable Energy Reviews*, **54**, 1668–1684.

Canadian GeoExchange Coalition (2010) The state of the Canadian geothermal heat pump industry: Industry Survey and Market Analysis. http://www.geo-exchange.ca/en/ UserAttachments/article64_Industry%20Survey%202010_FINAL_E.pdf (accessed July 26, 2016).

Florides, G. and Kalogirou, S. (2007) Ground heat exchangers - A review of systems, models and applications. *Renewable Energy*, **32** (**15**), 2461–2478.

Geothermal Heat Pump Consortium (2013) http://www.geoexchange.org (accessed January, 29, 2016).

Kulcar, B., Goricanec, D. and Krope, J. (2008) Economy of exploiting heat from low-temperature geothermal sources using a heat pump. *Energy and Building*, **40**, 323–329.

Omer, A.M. (2008) Ground-source heat pumps systems and applications. *Renewable and Sustainable Energy Reviews*, **12**, 344–371.

6

Design Considerations and Installation

General information on procedures before installation of ground-source heat pump (GSHP) systems such as ground thermal conductivity test is provided here. Furthermore, a short review of how the systems are designed based on the building loads and weather data is provided. Procedures in designing systems with unbalanced loads are also discussed.

6.1 Sensitivity to Ground Thermal Conductivity

Thermal characteristics of the ground play an important role in evaluating heat flow in the ground surrounding a borehole. One property that can affect the conduction of heat in the ground surrounding boreholes is ground thermal conductivity. Therefore, it is a key parameter in designing GSHP systems and seasonal borehole energy storage systems.

In order to understand the variation of temperature rise at the borehole wall with ground thermal conductivity, hypothetical ground types with given specific heat and density (1550 J/kg K and 1950 kg/m^3, respectively) and varied conductivities are considered in a heat injection/removal example in this section (Table 6.1). Note that the ground conductivity variation in this section is maintained within the range that often exists in the ground surrounding borehole heat exchangers. In the figures in this section, the temperature variation at certain points in the ground with various thermal conductivities is illustrated.

A two-dimensional model of transient heat conduction in the ground surrounding multiple ground heat exchangers (GHEs) is presented in this section. A domain consisting of several borehole systems, each consisting of 16 vertical boreholes is considered (Figure 6.1). The borehole systems are placed at every 100 m and the boreholes are installed at 6-m distances.

Due to the periodic seasonal changes in climate, the ground heat load profile for each borehole is expected to be periodic. Here, the ground load profile is modeled as a simplified sinusoidal profile delivering/removing a maximum of 30 W/m of heat (Figure 6.2). This heat flow rate per unit length of the borehole is within the range of the heat flow rates that are often chosen by system designers. Using this heat boundary at the borehole wall, the temperature of the ground surrounding the boreholes can be determined using a finite volume method. Details of this method and the geometry of the solution domain can be found in Section 8.3.

Geothermal Energy: Sustainable Heating and Cooling Using the Ground, First Edition.
Marc A. Rosen and Seama Koohi-Fayegh.
© 2017 John Wiley & Sons, Ltd. Published 2017 by John Wiley & Sons, Ltd.

Table 6.1 Ground thermal characteristics for four hypothetical compositions.

Property	Ground type 1	Ground type 2	Ground type 3	Ground type 4
Thermal conductivity, k (W/m K)	1.0	1.5	2	2.5
Specific heat (J/kg K)	1550	1550	1550	1550
Density (kg/m³)	1950	1950	1950	1950

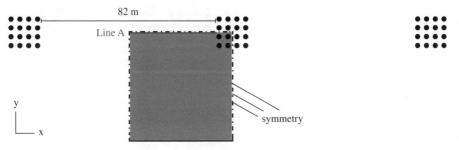

Figure 6.1 Solution domain for a region of ground containing several borehole systems.

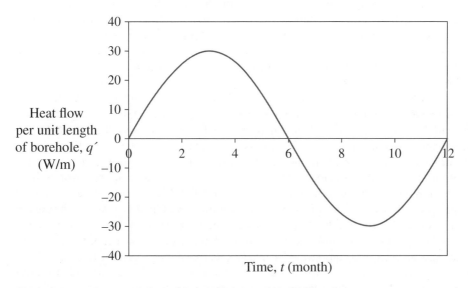

Figure 6.2 Heat flow rate at the borehole wall, per unit length of borehole.

The temperature at the wall of a borehole that is shown in Figure 6.3 is chosen as representative of the temperature of all boreholes in the borefield and its variation with time for various ground thermal conductivities is shown. It is seen that the temperature variations at the borehole wall are periodic. This is expected since the ground heat flow at the borehole wall varies periodically. For grounds with higher thermal conductivities, the temperature of the ground at the borehole wall is lower in the ground heat delivery

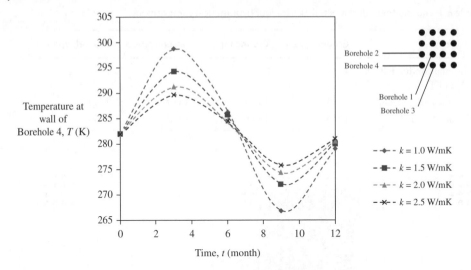

Figure 6.3 Variation of the temperature of the borehole wall with time, for several values of the ground thermal conductivity *k*.

mode and higher in the ground heat removal mode. This is due to the greater ability of grounds with higher thermal conductivity to conduct heat than to store it. An example of a ground with high thermal conductivity is one that contains pieces of rocks of high thermal conductivity. On the other hand, grounds containing air gaps are not good conductors of heat; therefore, a borehole installed in such ground types may experience a higher temperature at its wall in the heat delivery mode and a lower temperature at the borehole wall in the heat removal mode. For higher heat pump efficiencies, lower temperatures at the borehole wall in heat delivery mode and higher temperatures at the borehole wall in the heat removal mode are preferred. Therefore, a heat pump coupled with boreholes that are installed in grounds with a higher thermal conductivity are expected to have a higher efficiency. This is discussed in more detail in Section 11.3.1.

Figure 6.4 shows the temperature on an arbitrary line, Line A in Figure 6.1, after 3 and 9 months of system operation. These two times are chosen since the ground heat delivery/removal peaks at these times and it is expected that notable temperature rises occur at these times. By observing ground temperatures on Line A, the size of the area in the ground outside the borefield experiencing temperature rise can be determined. It is seen in Figure 6.4 that boreholes that are installed in grounds with higher thermal conductivities can cause a temperature rise in a larger area of the ground surrounding them during their yearly operation. This could mean that a borehole that is installed in a more conductive ground could have larger negative impacts on surrounding sensitive ecosystems than a borehole installed in ground with lower thermal conductivity. Observing the temperature contours in the ground surrounding the boreholes that are shown in Figure 6.5 and Figure 6.6 could give a better understanding of the above statements. It is seen that, although the thermal conductivity of the ground does have a notable effect on the temperature rise in the ground surrounding the borehole, the change in size of the area in the ground surrounding the boreholes that experience a temperature rise of over 0.1 K is less than 3 m in the current case. To complete this discussion, a long-term study of the system should be performed to examine the growth of the area in the ground

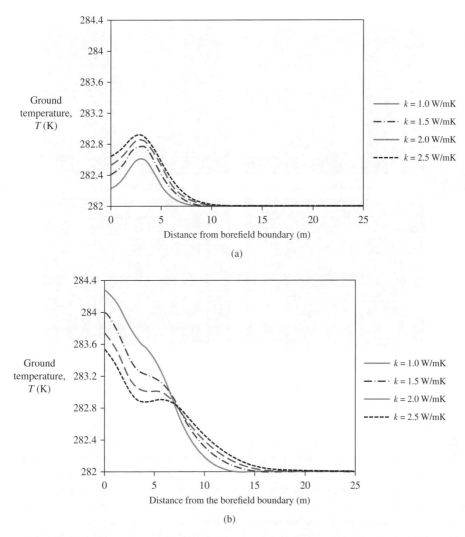

Figure 6.4 Ground temperatures outside of borefield boundary, for several values of ground thermal conductivity k, after (a) 3 months and (b) 9 months of system operation.

experiencing a temperature rise over several years. This is discussed in more detail in Section 10.3.

6.2 Thermal Response Test

As mentioned in the previous section, in designing GSHP systems and seasonal borehole energy storage systems, accurate information on thermal conductivity of the ground is important to avoid over prediction or under prediction of the system size. Since the ground thermal conductivity varies for various sites, a thermal conductivity test is usually performed to provide important information for the design of the GSHP system for a given site.

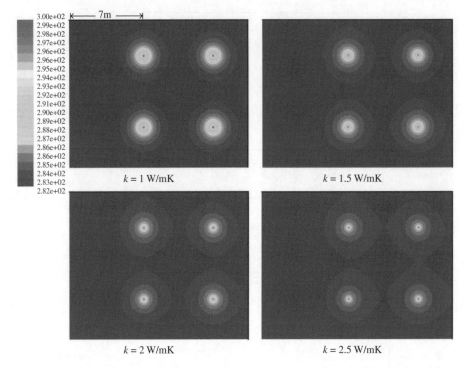

Figure 6.5 Ground temperatures contours after 3 months, for several values of ground thermal conductivity k.

The general method to measure ground thermal conductivity was first introduced by Mogensen (1983) and is referred to as the thermal response test (TRT). In principle, this method is similar to the needle probe test that is used to estimate the thermal conductivity of solids. The test has two main stages: the measurement test setup; and the mathematical modeling of the apparatus. The two stages are coupled using a recommended algorithm that uses the measured data and the mathematical model to estimate the thermal conductivity. In the measurement test, heat at constant power is injected into (or extracted from) a borehole by circulating a fluid in U-tubes. A mean fluid temperature is calculated over time by measuring and logging borehole fluid inlet and outlet temperatures. The measured data are compared against a mathematical model, such as the line-source theory, to calculate the thermal conductivity of the ground. The method also allows determination of borehole thermal resistance.

6.2.1 Test Setup

In setting up the test, parameters such as the initial ground temperature, borehole fluid flow rate, heat injection or extraction rate, and the test duration need to be known. The initial ground temperature may be measured using various proposed methods including temperature sensors, optical fibers and submersible wireless sensors. The operating fluid flow rate is often kept at the design ranges used when the borehole is in operation.

Martin and Kavanaugh (2002) suggest heating rates on the order of $50-80$ W/m. Uniform heat injection and extraction rates are often desired as they allow simpler analysis

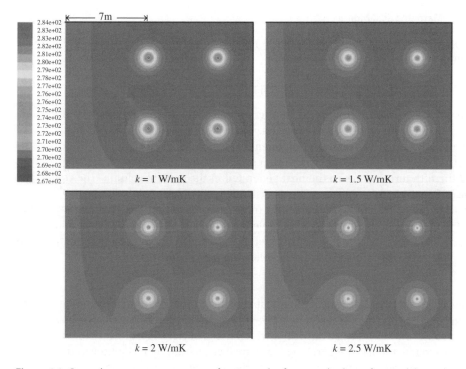

Figure 6.6 Ground temperatures contours after 9 months, for several values of ground thermal conductivity k.

procedures. Various designs have been proposed to achieve such a goal. For some applications, a combination of the designs can provide the desired results. Uniform heat input to the circulating fluid is achieved using electric resistance heaters immersed in the borehole fluid. However, in this method, the actual input to the ground can fluctuate due to changes in the heater voltage and heat losses or gains in the connecting piping. Improved insulation of the piping can help overcome the latter cause. In another method, a heat pump is used to provide heat injection and extraction. An active control valve coupled with temperature sensors at the inlet and outlet of the borehole fluid can eliminate much of the voltage fluctuations to provide uniform heat injection or extraction rates. The system, however, is costlier compared with electric heating.

The ground thermal conductivity and borehole thermal resistance often vary along the borehole depth. However, this variation is neglected in the conventional thermal response test where mean values for thermal conductivity of the surrounding ground and borehole resistance are estimated. To improve the test, a variety of analytical and numerical data analyses have been developed. For example, optical fiber technology can be applied to collect information about the temperature profiles in the borehole. While TRT logs the inlet and outlet fluid temperatures, the modified test measures borehole temperature along its length using fiber optic cables. This test is called the Distributed Thermal Response Test.

6.2.2 Mathematical Model

The general heat conduction equation in cylindrical coordinates appears in the following form:

$$\frac{\partial^2 T}{\partial r^2} + \frac{1}{r}\frac{\partial T}{\partial r} + \frac{1}{r^2}\frac{\partial^2 T}{\partial \varphi^2} + \frac{\partial^2 T}{\partial z^2} + \frac{\dot{q}_{gen}}{k} = \frac{1}{\alpha}\frac{\partial T}{\partial t} \tag{6.1}$$

where t is the time from the start of operation, α is the thermal diffusivity of ground, and T is the temperature of the ground. The first two terms on the left-hand side of Equation (6.1) are the heat flux components in the radial (r) direction, the third and the fourth terms are related to the circumferential (φ) and axial (z) directions, respectively, and the fifth term relates to the heat generated in the control volume. The right-hand side of Equation (6.1) represents the transient effects of heat conduction.

Various analytical and numerical heat models have been proposed for conduction of heat in the ground surrounding a borehole. Two analytical methods that are widely recognized and used are the line-source method and the cylindrical-source method. The principles of each are discussed in the following paragraphs.

6.2.2.1 Line-Source Model

The earliest approach to calculating heat transfer in the ground surrounding a vertical ground heat exchanger is Kelvin's line-source model, that is, the infinite line-source (Ingersoll et al. 1954; Hellström 1991) which uses Fourier's law of heat conduction. In the line-source theory, the borehole is assumed as an infinite line-source in the ground which is regarded as an infinite medium with an initial uniform temperature. Due to its minor order, heat transfer in the axial direction along the borehole, which accounts for the heat flux across the ground surface and down to the bottom of the borehole, can be neglected. This assumption is valid for a length of the borehole distant enough from the borehole top and bottom. Therefore, heat conduction in the ground is an unsteady radial heat conduction problem, that is, $T(r, t)$, and the following simplified heat conduction equation can be derived:

$$\frac{\partial^2 T}{\partial r^2} + \frac{1}{r}\frac{\partial T}{\partial r} = \frac{1}{a}\frac{\partial T}{\partial t} \tag{6.2}$$

The following assumptions are made for the line-source model in GHEs:

- Thermal properties of the ground are isotropic and uniform.
- Moisture migration is negligible.
- The impact of groundwater advection is negligible.
- Thermal contact resistance is negligible between the pipe and grout, the grout and the borehole wall, and between the borehole wall and the ground at the borehole wall.
- The effect of the ground surface is negligible.

The boundary conditions for a line source of heat are introduced as:

$$-2\pi r_b k \frac{\partial T}{\partial r} = q' \quad r \to 0$$

$$T - T_0 \to 0 \qquad r \to \infty$$

$$T - T_0 = 0 \qquad t = 0 \tag{6.3}$$

where T_0 is the initial temperature of the ground ($t = 0$), $q\prime$ is the heating rate per unit length of the line source, and k is the ground conductivity. The first boundary condition

in Equation (6.3) is related to the heat flow rate per unit length at the borehole wall conducted in the ground, which is derived from Fourier's law of heat conduction (Eskilson
1987). At larger distances ($r \rightarrow \infty$) the temperature of the ground is not affected by the
line source of heat and remains equal to the initial condition. The last condition relates
to the initial temperature of the ground at $t = 0$. The temperature response in the ground
due to a constant heat flow rate per unit length of the line source (q') is given by:

$$T(r, t) - T_0 = \frac{q'}{4\pi k} \int_{\frac{r^2}{4at}}^{\infty} \frac{e^{-u}}{u} du \qquad (6.4)$$

The left-hand side of Equation (6.4) gives the temperature excess of the ground surrounding a single borehole at radial distance r and at time t when heat flow rate per unit
length of the borehole (q') is transferred through the ground. The exponential integral
on the right-hand side of Equation (6.4) can be calculated numerically. It is seen that
a higher rate of heat flow (q') on the borehole wall results in a higher temperature rise
around the borehole.

The line-source model is simple and requires little computation time; it is therefore the
most widely used theory in design methods to analyze GHE heat transfer. However, due
to its assumption of the infinite line source, temperatures computed from this theory at
a short distance from the center and after a short time exceed the maximum possible
fluid temperature computed from an energy balance. Ingersoll and Plass (1948) estimate
that using this method may cause a noticeable error when $at/r_b^2 < 20$, where r_b is the
borehole radius, t is the time from the start of system operation, and α is the thermal
diffusivity of the ground surounding the borehole. Therefore, this method can only be
applied to small pipes for short-term operation of GSHP systems, that is, from a few
hours to months. To make the analytical results obtained by this method more accurate
and comparable with numerical ones, several studies have focused on improvements,
among which the results of Hart and Couvillion (1986) are some of the most accurate.
They propose an equation for the ground temperature around a line source in terms
of a power series of the ratio of radial distance and farfield distance. The definition
of farfield distance depends on the radius of the borehole. Lamarche and Beauchamp
(2007) develop alternative forms for the finite line-source solution with shorter computation times.

The exponential integral in Equation (6.4) has been approximated in various ways. For
example, Carslaw and Jaeger (1946) use the following approximation:

$$\int_{\frac{r^2}{4at}}^{\infty} \frac{e^{-u}}{u} du = \ln\left(\frac{4\alpha t}{r^2}\right) - \gamma \qquad (6.5)$$

where γ (Euler's constant) = 0.5772. The error related to this approximation is less than
2% when $t > 5r^2/\alpha$.

The effect of the thermal resistance between the borehole fluid and the borehole wall
is accounted for by assuming a borehole resistance as follows:

$$T_f(t) = T_b(t) + q \cdot R_b \qquad (6.6)$$

Applying the approximation in Equation (6.5) and accounting for the borehole thermal
resistance, the borehole fluid temperature can be written in relation to the heat flow rate,
the thermal conductivity of the infinite medium (i.e., the ground), the borehole thermal

resistance, and time as follows:

$$T_f(t) = T_0 + \frac{q'}{4\pi k}\left[\ln\left(\frac{4\alpha t}{r^2}\right) - \gamma\right] + q \cdot R_b \tag{6.7}$$

Further simplification of Equation (6.7) results in the following:

$$T_f(t) = A\ln t + B \tag{6.8}$$

where

$$A = \frac{q'}{4\pi k}, \quad B = T_0 + \frac{q'}{4\pi k}\left[\ln\left(\frac{4\alpha}{r^2}\right) - \gamma\right] + q \cdot R_b$$

If the heat input is maintained at a constant rate, the variation of the borehole fluid temperature with time can help predict the thermal conductivity of the ground using Equation (6.8). This is the basis of a thermal response test using measured data and line-source theory to estimate ground thermal conductivity. Further information on this topic can be found elsewhere (Sanner *et al.* 2005; Raymond *et al.* 2011; Wagner *et al.* 2012). Since it is difficult to maintain a constant power supply and heating rate during the test, further relations have been derived to improve Equation (6.8) so it can account for a varying heat flow rate during the test. These include the use of stepwise constant heat pulses that can be superposed in time to account for such variations. For brevity, these derivations are not included in this section, but details can be found elsewhere (Yavuzturk 1999; Yavuzturk and Spitler 1999), and the ideas are revisited in Section 7.2.2.3.

6.2.2.2 Cylindrical-Source Model

Another analytical model based on Fourier's law of heat conduction was first developed by Carslaw and Jaeger (1946). In this model, the borehole is assumed to be a cylindrical pipe with infinite length buried in the ground, which is considered a homogeneous infinite medium with constant properties. During the transient stage of heat storage in the ground, the thermal capacities of the fluid and immediate region next to the core are neglected in the early time results of the cylindrical-source theory. In addition, it is assumed that heat transfer between the borehole and ground with perfect contact is pure heat conduction. Therefore, using the same assumptions presented for the line-source theory in the previous section, the governing equation of the transient heat conduction in cylindrical coordinates can be simplified as:

$$\frac{\partial^2 T}{\partial r^2} + \frac{1}{r}\frac{\partial T}{\partial r} = \frac{1}{a}\frac{\partial T}{\partial t} r_b < r < \infty \tag{6.9}$$

with the following boundary conditions:

$$-2\pi r_b k \frac{\partial T}{\partial r} = q' \qquad r = r_b,\ t > 0$$
$$T - T_0 = 0 \qquad r > r_b,\ t = 0 \tag{6.10}$$

where r_b is the borehole radius and T_0 is the initial temperature of the ground. The governing equation for this model can be solved analytically for either a constant pipe surface temperature or a constant heat transfer rate from the pipe to the ground. The analytical solution of Equation (6.9) given by Carslaw and Jaeger (1946) is:

$$T(r,t) - T_0 = \frac{q'}{k}G\left(\frac{at}{r_b},\frac{r}{r_b}\right) \tag{6.11}$$

and $G(z, p)$ is a function of time (t) and distance from the borehole center (r), and involves integrations from zero to infinity of a complicated function, including Bessel functions (Bandyopadhyay *et al.* 2008):

$$T(r,t) - T_0 = \frac{q'}{\pi^2 k} \int_0^\infty \frac{e^{-\frac{uat}{r^2}} - 1}{J_1^2(u) + Y_1^2(u)} [J_0(u)Y_1(u) - J_1(u)Y_0(u)] \frac{du}{u^2} \tag{6.12}$$

To obtain the temperature on the borehole wall ($r/r_b = 1$), which is the representative temperature in the design of GHEs, some graphical results and tabulated values for the $G(at/r_b, r/r_b)$ function at $r/r_b = 1$ (the borehole wall) can be found in related references (e.g., Ingersoll *et al.* 1954).

Using the cylindrical-source method in the thermal response test mathematical analysis, the effect of the thermal resistance between the borehole fluid and the borehole wall is accounted for by assuming a borehole resistance [Equation (6.6)] and borehole temperature can be derived as:

$$T_f(t) = T_0 + \frac{q'}{k} G\left(\frac{at}{r_b}, \frac{r}{r_b}\right) + q \cdot R_b \tag{6.13}$$

Equation (6.13) is comparable with Equation (6.7) when the line-source theory is used to relate the borehole temperature to the heat flow rate and ground thermal conductivity. Similar to the line-source model, cylindrical-source solutions have limitations at the early stage of transient heat conduction flux build up after a step heat input is applied to the system fluid. The cylindrical-source solution assumes a steady flux across a hollow cylindrical surface (borehole boundary) and omits the grout and fluid from the problem domain. Yet during the transient flux build up, the thermal capacities of the fluid and immediate region next to the core are neglected in the early time results of the cylindrical-source theory.

Kavanaugh (1992) moved the reference cylindrical surface from the borehole boundary to an intermediate surface inside the borehole nearer to the core to improve the accuracy of the cylindrical-source solution. This modification allows the reference surface to reach a near-steady-flux condition earlier than with the borehole boundary as the reference surface. However, the effect of neglecting the thermal capacity of the fluid remains a shortcoming of the cylindrical-source theory.

Hellström (1991) applies a numerical inversion technique to solve the inverse Laplace transform of the governing differential equation for the one-dimensional transient heat conduction equation in polar coordinates and develops an alternative form for the cylindrical-source solution.

Hikari *et al.* (2004) derive simplified forms for the cylindrical-source solution at the borehole surface depending on the Fourier number.

6.3 Building Energy Calculations

One of the first steps in designing any heating and cooling system for a space is the calculation of heat loss and gain in the heating and cooling seasons, respectively. In the heating season, the calculation of total heat loss includes the sensible heat loss through conduction, infiltration, and ventilation loads. In the cooling season, the calculation of total heat gain includes conduction, infiltration, ventilation, and radiation. Accurate load calculations have a direct impact on energy efficiency of the heating system. The follow-ing are some of the parameters that need to be defined in order to accurately calculate heating and cooling loads:

- location (e.g., latitude, elevation, and weather data)
- building orientation (for passive solar heat gain calculations)
- internal temperature and humidity requirements
- building user habits
- internal heat gains (e.g., occupants, lights, electronics and appliances)
- building enclosure (i.e., building air tightness)
- building construction (walls and window conductivities)

While some building energy calculations are aimed at sizing heating systems under a set of typical conditions for the above parameters, building energy calculations can involve estimation of these parameters over time (e.g., in hourly or shorter time steps). For example, instead of using a typical winter temperature for a location in energy calcu-lations, monthly, daily or even hourly temperature variations of the air can be used. The latter energy calculations are often part of a more thorough building simulation often for purposes other than sizing the heating and cooling system, for example, building per-formance analysis. The level of accuracy of calculations of temperatures, energy flows and energy consumption depends on the building type and the purpose of the calcula-tions. In the next few subsections, information on some of the parameters that affect energy calculations is provided. References are provided for more detailed information on every topic.

As mentioned in Sections 3.2 and 5.1, if the ground temperatures in the heat injection or removal are varied, the performance of the GSHP could deteriorate. In order for the GSHP system to operate sustainably, ground heating and cooling loads should be some-what balanced to ensure the temperature of the ground does not progressively increase or decrease as a result of over-heating or over-cooling. In Europe, where heating loads are often higher than cooling loads, GSHPs are used to cover base heating loads and are integrated with a supplementary heating system that covers peak heating loads. Basic heating systems include those based on fossil fuels and electricity. The use of renewable energy sources with a heat pump cycle is an additional option for handling unbalanced loads in regions with larger heating loads. In colder climates, where the cooling load is lower than the heating load, the heat pump can be integrated with solar thermal systems to balance the heat that is delivered to the ground in the cooling season with that to be extracted in the heating season.

6.3.1 Weather Data

One of the main parameters in building energy calculations is the collection of weather data. The outdoor temperatures for various locations are often measured and are avail-able in the tabulated formats. A common data type available is the dry-bulb temperature of the air on an hourly or daily basis. More generally, weather averages are used to sum-marize or describe the average climatic conditions of various locations. For example, the temperature averages and extremes for Canadian locations are available in Envi-ronment Canada (2016). Average monthly air temperatures throughout the year for Toronto, Ontario based on Canadian climate station data from 1971 to 2000 are shown

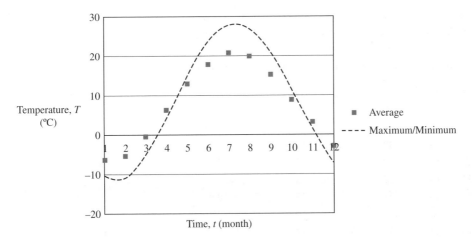

Figure 6.7 Average monthly air temperatures throughout the year for Toronto, Ontario, Canada.

in Figure 6.7. It is seen that the heating and cooling load profile throughout the year varies depending on the outside air temperature.

As noted in the previous section, the objective of a particular building energy analysis dictates what detailed weather information is needed. The format in which weather data is provided is often adjusted to a certain energy calculation method. For example, in one building energy calculation method, weather data are required in the form of temperature range bins (e.g., 5-Fahrenheit-degree) with the hours of occurrence for each bin. This type of weather data is used in methods assuming that all hours during a year when a particular temperature occurs can be grouped together and used in load calculations once, to avoid repetition of load calculations every time a certain temperature occurs. The numbers of hours that temperatures occur in 23 2.8-Celsius-degree (5-Fahrenheit-degree) bins in each year in Belleville, IL, USA are shown in Table 6.2 (USAF 1978). A more detailed building energy analysis using such a method requires monthly weather data (Table 6.3). These values can be used to predict energy consumption patterns for cooling and heating equipment at different times of the year.

6.3.2 Building Considerations

In this section, the weather-related annual periodic variations with the heat flux at the wall of the heat exchanger (here, the borehole) are correlated using building and weather specifications for a case study.

A building in Belleville, IL, USA is considered. The simplified load profiles as shown in Figure 6.8 are given by:

Heating load:

$$\dot{q}_{HL} = 32.7 - 2.7T_o \tag{6.14}$$

Cooling load:

$$\dot{q}_{CL} = 2.7T_o - 52.3 \tag{6.15}$$

This correlation yields $\dot{q}$ in units of kilowatt and requires that the temperature be in units of degrees Celsius. Note that these load profiles are assumptions for an arbitrary

Table 6.2 Dry-bulb temperature hours for an average year in Scott AFB, Belleville, IL, USA; period of record = 1967 to 1996.

Temperature range (°C)			Bin temperature (°C)	Total hours
38	to	40	39	3
35	to	37	36	37
32	to	34	33	160
29	to	32	31	325
27	to	29	28	509
24	to	26	25	690
21	to	23	22	848
18	to	21	19	770
16	to	18	17	730
13	to	15	14	654
10	to	12	11	602
7	to	9	8	603
4	to	7	6	631
2	to	4	3	667
−1	to	1	0	596
−4	to	−2	−3	379
−7	to	−4	−6	222
−9	to	−7	−8	145
−12	to	−10	−11	95
−15	to	−13	−14	51
−18	to	−16	−17	25
−21	to	−18	−19	10
−23	to	−21	−22	6
−26	to	−24	−25	2
−29	to	−27	−28	0

Source: Adapted from USAF (1978).

building. It is assumed that the building does not have any shift-breakdowns, that is, the building is used in the same way during a 24-h period. In other words, we assume the internal heat gains by people, equipment, lights, and so on do not vary temporally during the day. However, in the case of an office building, for example, the building is used for only 8 h a day resulting in heating and cooling load profiles for the 8-h period and when the building is not occupied. Due to different internal heat gains and sometimes a different thermostat temperature setting, the heating and cooling loads of the building exhibit a different profile. Note that the balance point for this building is approximately 13 °C (55 °F) and 19 °C (67 °F) for heating and cooling modes, respectively.

6.3.3 Heat Pump Considerations

Factors related to the performance of the heat pump unit [such as heating and cooling capacities, coefficient of performance (COP), power consumption, water pressure

Table 6.3 Dry-bulb monthly temperature hours for an average year in Scott AFB, Belleville, IL, USA; period of record = 1967 to 1996.

Temperature (°C)	Jan.	Feb.	Mar.	Apr.	May	Jun.	Jul.	Aug.	Sep.	Oct.	Nov.	Dec.
39							2	1				
36							18	12	2			
33						1	64	43	12	0		
31			0		0	5	102	83	34	4		
28			3	1	4	35	138	118	64	14	0	
25		1	7	4	22	76	163	153	94	33	3	
22	1	2	17	16	47	109	149	160	137	57	11	1
19	2	5	30	29	75	134	71	99	120	91	26	4
17	6	13	47	52	110	153	29	52	103	112	53	13
14	13	24	65	78	140	105	8	19	77	121	65	26
11	22	31	87	105	134	64	1	4	47	113	84	36
8	35	50	105	109	100	28		0	21	95	105	58
6	72	84	111	107	63	7			7	59	119	102
3	112	119	108	97	36	1			2	33	117	140
0	143	111	89	66	10				0	11	82	140
−3	102	90	50	34	3					2	35	96
−6	84	54	16	18	0						14	54
−8	64	37	5	3							5	33
−11	41	29	2	0							2	20
−14	26	15	1								0	10
−17	13	6	0									6
−19	5	1										3
−22	2	1										2
−25	1	0										1
−28	0											

Adapted from USAF (1978).

drop at the water-to-refrigerant heat exchanger] vary with ground temperature and the building load. Therefore, bin summaries are needed to calculate the heat pump electrical power use as well as its capacity for several values of ground temperature throughout the year. Table 6.4 shows an example of heat pump performance variation with ground temperature.

The numbers of hours that temperatures occur in 23 2.8-Celsius-degree (5-Fahrenheit-degree) bins for each month in Belleville, IL, USA (Table 6.3) are used as an example in order to predict energy consumption patterns for cooling and heating equipment at different times of the year (USAF 1978). Using the load profiles [Equation (6.12)] and the heat pump performance, the bin calculation procedure is performed. Note that the heat pump integrated capacity and the rated electric input are calculated separately for each month and vary for other months based on the average temperature of that

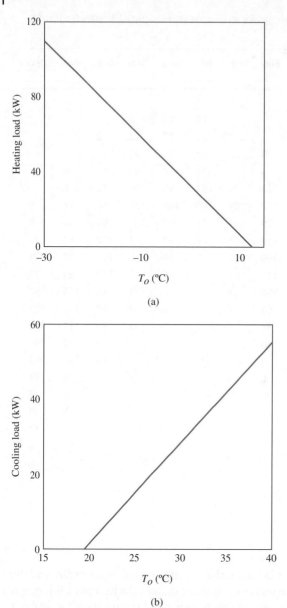

Figure 6.8 Heating and cooling loads for a building in Belleville, IL, USA. (a) Heating load profile, (b) cooling load profile.

month, as seen in Figure 6.9. These average temperatures are obtained with an iterative procedure assuming a transient profile for ground temperature and correcting it to the results after the first-year simulation until the assumption for the ground temperature leads to the same ground temperature in the simulation.

6.3.4   Load Calculations

Using heating and cooling calculation methods (e.g., the bin method), the variation of heating and cooling load profiles for the building throughout the year can be

Table 6.4 Typical heat pump heating and cooling capacities at an air flow rate of 6000 CFM (2.8 m³/s).

Ground temperature (°C)	Heating capacity (kW)	Total power input (kW)	Cooling capacity (kW)	Total power input (kW)
	At indoor dry bulb temperature 21 °C		At indoor dry bulb temperature 24 °C	
3	39.9	16.6	57.8	18.9
6	45.1	17.4	56.3	19.2
8	50.4	18.3	54.9	19.6
11	53.9	18.8	53.4	19.9
14	57.4	19.3	51.9	20.3
17	61.0	19.9	50.5	20.6

determined. For example, performing bin calculations similar to the one shown in Table 6.5, the variation of heating and cooling loads of the building throughout the year are determined (Figure 6.10). In a balanced system, almost all the heat that is stored in the ground during the summer, is used in the winter. However, many parts of the world do not have a balanced climate. Furthermore, additional heating systems may be integrated to the GSHP system which affects their design. Some GSHP systems are designed to cover the peak cooling load and are, therefore, oversized for heating. Some are designed based on the base heating load and integrate another heater to the system that operates using fossil fuels. It is seen in Figure 6.10 that the heating and cooling loads of the building example from the previous section are not balanced throughout the year; there are 8 months of heating and 4 months of cooling. In order to balance the amount of heat that is stored in the ground, the size of the GHE is designed based on the cooling load, and supplemental heat in the form of electrical resistance heating is required when the heating load from the GHE is not met.

6.3.5 Ground Heat Injection and Extraction

The average temperature of the ground depends on how the cycle of the heat storage removal starts the first time the system starts to operate. It would become lower or higher than the initial temperature of the ground according to the first operation time in winter or summer. Here, we consider that the heat is stored in the ground during the 4-month cooling season through two vertical boreholes of 200 m length. Based on the building heating and cooling loads calculated from the bin data, the magnitude of the heat flux from the borehole wall is determined (Figure 6.11). It is seen there that a sinusoidal function can be fit to the monthly cooling load data. Note that due to the unbalanced weather of this area, the heat that is stored in the ground in the cooling season only covers part of the total heating load in the 8-month heating season; therefore, in the heating months a curve must be chosen that results in the same amount of heat removal from the ground over its period as the heat stored in the ground in the cooling

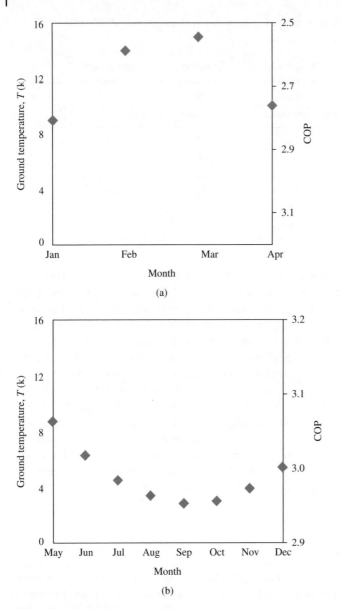

Figure 6.9 Transient ground temperature in response to (a) heat injection and (b) heat extraction. COP, coefficient of performance.

season. That is,

$$\int_0^{day_{sr}} q''_{storage}\, d(day) + R_s \int_{day_{sr}}^{365} q''_{removal}\, d(day) = 0 \qquad (6.16)$$

where R_s is the ratio of heat extraction to heat injection for the ground. This parameter accounts for part of the stored heat in the cooling mode that passes away from

Table 6.5 Sample of bin energy calculations for the month of July in Scott AFB, Belleville, IL, USA.

	Weather		Building				Heat pump				
Temperature	Temperature difference, $T_{bin} - T_{bal}$	Hours in July	Building cooling load	Heat pump integrated capacity	Cycling capacity adjustment factor	Adjusted heat pump capacity	Operating time fraction	Rated electric input	Seasonal heat pump electric consumption	Heat exchanger heat injection
(°C)	(°C)	(h)	(kW)	(kW)		(kW)		(kW)	(kWh)	(kWh)
39	17.9	2	52.3	51.9	1.00	52.0	1.00	20.3	40.6	145.2
36	15.1	18	44.8	51.9	0.97	50.1	0.89	20.3	326.9	1134.0
33	12.3	64	37.4	51.9	0.93	48.3	0.77	20.3	1006.2	3397.6
31	9.6	102	29.9	51.9	0.89	46.4	0.64	20.3	1334.6	4383.7
28	6.8	138	22.4	51.9	0.86	44.5	0.50	20.3	1411.0	4505.0
25	4.0	163	14.9	51.9	0.82	42.7	0.35	20.3	1159.8	3596.1
22	1.2	149	7.5	51.9	0.79	40.8	0.18	20.3	554.4	1667.9
								Total	5833.4	18829.5

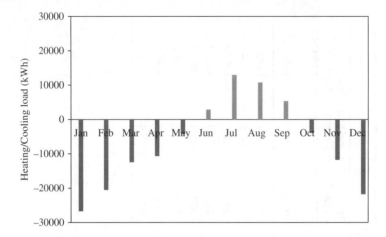

Figure 6.10 Variation of heating and cooling loads of the building throughout the year. Note that here positive values represent cooling loads and negative values represent heating loads (i.e., negative cooling loads).

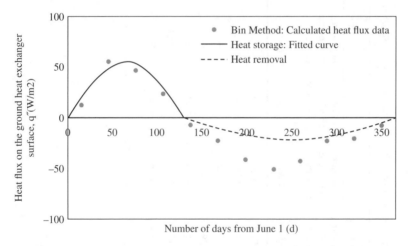

Figure 6.11 Variation of heat flux on the ground heat exchanger wall with time for 1 year.

the borefield and cannot be extracted in the heating mode. Starting from the first day of heat storage in the ground, day_{sr} is the last day of heat injection and the start of heat removal from the ground. Based on Equation (6.16), a sinusoidal curve can be chosen that represents the heat injection or extraction profile based on the provided data from the load analysis of the building (solid lines in Figure 6.12), and the other mode of operation (heat extraction or injection) can be defined based on Equation (6.16) that results in a balanced system that collects all the heat that it injects into the ground (dotted lines in Figure 6.12). Three typical heat injection and extraction profiles are shown in Figure 6.12 where a naturally balanced profile (6 months heat injection and 6 months heat extraction) is compared with two systems with unbalanced heat injection and extraction needs, where one mode of heat injection or extraction is balanced according to Equation (6.16) (dotted lines).

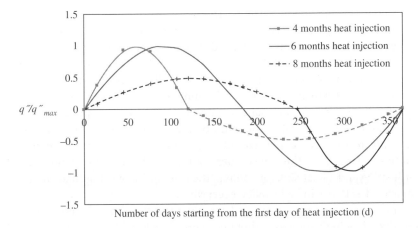

Figure 6.12 Typical balanced heat injection and extraction temporal profiles for 1 year.

The following heat flux profiles for heat injection and extraction are chosen for the current unbalanced-load system where the system injects heat into the ground for 130 days. The amplitude of the heat extraction profile is chosen so that the system would remain balanced [Equation (6.18)]. Substituting the number of days (*day*) from June (start of cooling season), this correlation yields q'' in units of kilowatt (Figure 6.11):

Heat storage:

$$q''(d) = 55.9 \, \sin \left(\frac{day}{130}\pi \right) \tag{6.17}$$

Heat removal:

$$q''(d) = -0.70 \times 53.7 \, \sin \left(\frac{day - 130}{235}\pi \right) \tag{6.18}$$

The amplitude of the sinusoidal heat injection profile in Equation (6.17) is chosen assuming an average unit heat injection rate of 11 W/m. Note that the ratio of heat extraction to heat injection (R_s) in the above annual profile is chosen based on an iterative procedure. At $R_s = 0.7$, the system appears to have collected all the heat that it stored in the ground. To model the heat exchanger system and assess the effects of a periodic variation of heat flux on the borehole wall, we use the fitted curve for the cooling season in Figure 6.11 and the curve for heat removal from the ground resulting in the same amount of heat removal from the ground. Equation (6.17) and Equation (6.18) can be used as boundary conditions on the borehole wall.

6.4 Economics

About one-third of the world population has poor living standards and no or little access to the energy services typical of modern life in economically developed countries. With a continually increasing world population, a key issue for improving living standards is to make energy available to all at affordable prices. As long-term prices of fossil fuels continue to rise, the costs related to geothermal energy systems are becoming more competitive with conventional energy systems. Therefore, geothermal energy systems can potentially provide energy at lower costs than conventional energy systems. Economic analyses of geothermal systems are needed to gain support from governments and industries for their future use and development. Some of the approaches used to determine the economic feasibility of GSHP systems are as follows: the net present value method; the internal rate of return method; and the annual cost method.

The economic feasibility of a geothermal system varies according to type and application. Initially, systems for electricity generation or direct use were considered economically viable only in areas where thermal water or steam was found concentrated at depths close to the surface. With the development of GSHPs that use the ground as a heat source for heating and a heat sink for cooling, the economics of using such energy systems has improved as they can be installed everywhere.

When calculating the economic savings related to the installation of geothermal energy systems, the following parameters should be considered:

- **Local energy costs.** In countries where the price of fossil fuels or electricity is low, the operating costs of conventional heating systems are relatively low. In such cases, a longer time period is needed before the investment on the costly installation is returned.
- **Interior temperature, local weather, and ground heat exchanger configuration.** Such parameters affect the COP of the heat pump. In more extreme climates with very low winter temperatures, the heat pump will operate at lower COPs and its operating costs can end up being higher, depending on the source and method used for electricity generation.
- **Required energy form for the GSHP system to operate.** The use of GSHPs is more economically feasible in countries that fully or partially generate electricity from sources other than fossil fuels, such as renewable and nuclear, as the cost of such electricity generation is lower than the cost of electricity generated from fossil fuels, resulting in lower operating costs for GSHP operation.

GSHP systems tend to be economically beneficial in the long run (i.e., over their lifetimes) due to their low operating costs (e.g., no fuel costs compared with fossil-fuel heating systems). Their installation costs are often higher than conventional fossil-fuel-based system heaters. However, more research and development and advances in the design of geothermal systems could lower their costs. Furthermore, they can be more economically beneficial with increased financial incentive schemes, commonly funded by governments and electric utilities, for environmentally favorable options such as heat pumps. Further geothermal systems can generate cost savings related to CO_2 taxes. Since they provide heating and cooling with a single installation, their costs are often comparable with any type of combined heating and cooling system and they are, therefore, used in applications requiring both heating and cooling.

Several examples are provided in the literature on the economics of GSHP systems, and how they compare to other heating and cooling options. In the following paragraphs, the cost of a geothermal system and the parameters on which it depends are discussed through two examples.

6.4.1 Economic Analysis of a Ground-Source Heat Pump for Heating and Cooling a Single Building

Leong *et al.* (2011) simulate an existing 227 m^2 (2440 ft^2) old house located in Wasaga Beach, Ontario, Canada. The house is equipped with electric resistance heating, serviced through forced air and baseboard convection units. The annual energy use for electric heating is 30.63 MWh. If an optimum-design GSHP system with a double-layer horizontal ground heat exchanger (HGHE) of 500 m at 1 and 1.5 m depths is in operation, the electrical energy consumption for heating would be 10.89 MWh (from Table 8.4 for Case BC), which yields a electricity savings of 19.74 MWh. Here, we consider a rate for electricity of \$0.12/kWh (including delivery, regulatory, debt retirement, loss factor adjustment, and tax charges), which is typical of the residential rate for electricity in Ontario in recent years. Therefore, the monetary savings is \$2369/year in heating costs, where monetary values used here are 2016 Canadian dollars.

Assuming a fixed annual interest rate of 2%, the payback periods calculated using a present worth – annual payment analysis for the following initial costs are:

- \$10 000: 4.5 years
- \$15 000: 6.8 years
- \$20 000: 9.3 years

A typical GSHP system can be in service for at least 20 years. After the payback period, the \$2369/year in savings is available to the system owner.

This economic analysis is done without considering system cooling costs because the house is equipped with two window-mounted air conditioners for cooling and it is difficult to compare a central system to such air conditioners. However, because of its high efficiency one would expect an even shorter payback period if cooling is included in the analysis.

Environmental benefits are also important. If natural gas, with a heating value of 10.35 kWh/m^3 and a total cost of \$0.48/m^3, is to be used for heating, the following two scenarios can be analyzed:

- If a natural gas furnace with an efficiency factor of 0.8 (including the effects of rated full-load efficiency, part-load performance and oversizing) is installed to heat the house, the annual amount of natural gas use is 3700 m^3 at \$1776. The annual cost is about \$470 more than operating the GSHP system.
- If the electricity for the GSHP system is generated using natural gas and delivered with an overall efficiency of 30%, the annual amount of natural gas use is 3510 m^3, which is about 190 m^3 less than the amount required by a natural gas furnace.

These two scenarios together show the economic and environmental benefits of operating the GSHP system, that is, an annual savings of \$470 relative to operating a natural gas furnace for heating.

6.4.2 Comparison of Economics of a Ground-Source Heat Pump and an Air-Source Heat Pump

Esen *et al.* (2007) compare GSHP and air-source heat pump (ASHP) systems techno-economically. The systems are designed for space cooling and are connected to a test room in Firat University, Elazig, Turkey. The comparison is based on experimental performance evaluations of the GSHP and ASHP systems.

The ASHP system contains two circuits, one being a refrigerant (R-22) circuit and the other an air circuit (i.e., the condenser fan circuit). The GSHP system mainly has three circuits:

- a refrigerant circuit;
- a fan coil (or air) circuit;
- a GHE circuit, in which a water–antifreeze solution flows. This circuit includes a 100 m horizontal length of pipe divided into two GHEs. Each GHE has a 50 m length, 0.3 m of pipe separation distance, and a 0.016 m nominal pipe diameter. Each GHE also has 15 m^2 spacing where buried in trenches of 1 and 2 m depths, and they are referred to here as horizontal ground heat exchanger (HGHE1) and horizontal ground heat exchanger (HGHE2).

To assess the annual performance of the GSHP and ASHP systems, the annual mean air and ground temperatures are required as boundary conditions. These temperatures, based on experimental data, are used, along with the mean ground temperatures in winter and summer. The experimental results were obtained from June to September in the cooling season of 2004. The average cooling coefficient of performance (COP$_{sys}$) of the GSHP system for the horizontal ground heat exchanger (HGHE) in trenches of 1 and 2 m depths are 3.85 and 4.26, respectively, and COP$_{sys}$ for the ASHP system is 3.17.

The annual cost method is employed to assess and contrast the economics of the GSHP and ASHP systems. Using a lifetime of 20 years for both systems, an interest rate of 8% and an annual fuel price escalation rate of 4%, the annualized life cycle cost has been calculated for each cooling system based on current electricity prices in Turkey. To estimate the GSHP and ASHP operating expenditures, the systems are assumed to operate 12 h per day on average for 108 days (June 1 to September 16, 2004). For this period, the price of the electricity consumed for the cooling load of the test room is found for the GSHP-1m, the GSHP-2m and the ASHP systems to be 78, 70 and 95 Euro/season, respectively. Correspondingly, the payback periods for the GSHP-1m and GSHP-2m systems are 3.7 and 4.0 years when compared with the ASHP system. The operating cost of each GSHP system is less than that for the ASHP system, but the capital cost of the ASHP system is less than that for both GSHP systems.

6.5 Standards

A number of GSHP standards and guidelines are available to help ensure quality and to promote standardization. This is particularly important in the design and installation of GSHP systems.

For instance, the following binational standard was recently published: ANSI/CSA C448 Series-16 Design and installation of ground-source pump systems for commercial and residential buildings (CSA Group 2016). Following a performance-based approach and best practices, the Standard sets requirements in areas such as equipment and material selection, design, commissioning, and decommissioning. The Standard has several objectives:

- harmonize differences between existing GSHP resources;
- simplify referencing in GSHP regulations and contracts;
- assist in developing GSHP contracts;
- incorporate the latest advances into GSHP and related technology;
- clarify compliance requiements;
- provide and enhance credibility through an accredited and neutral standards development process.

The Standard was developed by experts from across North America, including leaders from industry, trade and professional associations, utilities, drillers, installers, manufacturers, regulators, designers, engineers, researchers, and academics.

The ANSI/CSA C448 has several parts, covering various types of GSHPs and various aspects of their use:

- ANSI/CSA C448.0: Design and installation of ground-source heat pump systems – Generic applications for all systems
- ANSI/CSA C448.1: Design and installation of ground-source heat pump systems for commercial and institutional buildings
- ANSI/CSA C448.2: Design and installation of ground-source heat pump systems for residential and other small buildings
- ANSI/CSA C448.3: Installation of vertical configured closed-loop ground-source heat pump systems
- ANSI/CSA C448.4: Installation of horizontal configured closed-loop ground-source heat pump systems
- ANSI/CSA C448.5: Installation of surface water (including submerged exchangers) heat pump systems
- ANSI/CSA C448.6: Installation of open-loop systems ground water heat pump systems
- ANSI/CSA C448.7: Installation of standing column well heat pump systems
- ANSI/CSA C448.8: Installation of direct expansion heat pump systems

References

Bandyopadhyay, G., Gosnold, W. and Mannc, M. (2008) Analytical and semi-analytical solutions for short-time transient response of ground heat exchangers. *Energy and Buildings*, **40** (**10**), 1816–1824.

Canadian Standards Association (CSA) Group (2016) Design and installation of ground source heat pump systems for commercial and residential buildings, ANSI/CSA C448 Series-16.

Carslaw, H.S. and Jaeger, J.C. (1946) *Conduction of Heat in Solids*, Claremore Press, Oxford.

Environment Canada (2016) Canadian Climate Normals. http://climate.weather.gc.ca/climate_normals/index_e.html (accessed January 29, 2016).

Esen, H., Inalli, M. and Esen, M. (2007) A techno-economic comparison of ground-coupled and air-coupled heat pump system for space cooling. *Building and Environment*, **42** (**5**), 1955–1965.

Eskilson, P. (1987) *Thermal analysis of heat extraction boreholes*. PhD thesis. University of Lund.

Hart, D.P. and Couvillion, R. (1986) *Earth Coupled Heat Transfer*, National Water Well Association, Dublin, OH.

Hellström, G. (1991) *Ground heat storage: Thermal analyses of duct storage systems*. PhD thesis. University of Lund.

Hikari, F., Ryuichi, I. and Takashi, I. (2004) Improvements on analytical modeling for vertical U-tube ground heat exchangers. *Geothermal Resources Council Transactions*, **28**, 73–77.

Ingersoll, L.R. and Plass, H.J. (1948) Theory of the ground pipe heat source for the heat pump. *ASHVE Transactions*, **47**, 339–348.

Ingersoll, L.R., Zobel, O.J. and Ingersoll, A.C. (1954) *Heat Conduction with Engineering, Geological, and other Applications*, McGraw-Hill, New York.

Kavanaugh, S.P. (1992) Simulation of ground-coupled heat pumps with an analytical solution. *Proceedings of the ASME International Solar Energy Conference*, **1**, 395–400.

Lamarche, L. and Beauchamp, B. (2007) A new contribution to the finite line source model for geothermal boreholes. *Energy and Building*, **39** (2), 188–198.

Leong, W.H., Tarnawski, V.R., Koohi-Fayegh, S. and Rosen, M.A. (2011) Ground thermal energy storage for building heating and cooling, in *Energy Storage* (ed. M.A. Rosen), Nova Science Publishers, New York, pp. 421–440.

Martin, C.A. and Kavanaugh, S.P. (2002) Ground thermal conductivity testing – controlled site analysis. *ASHRAE Transactions*, **108**, 945–952.

Mogensen, P. (1983) Fluid to duct wall heat transfer in duct system heat storages. *International Conference on Subsurface Heat Storage in Theory and Practice*, June 6–8, 1983, Stockholm, Sweden. Swedish Council for Building Research, pp. 652-657.

Raymond, J., Therrien, R., Gosselin, L. and Lefebvre, R. (2011) A review of thermal response test analysis using pumping test concepts. *Ground Water*, **49**, 932–945.

Sanner, B., Hellström, G., Spitler, J., and Gehlin, S. (2005) Thermal response test – current status and world-wide application. *Proceedings of World Geothermal Congress 2005*, April 24–29, 2005, Antalya, Turkey. http://sanner-online.de/media/1436.pdf (accessed March 10, 2016).

USAF (1978) *Engineering Weather Data*, Department of the Air Force Manual AFM 88-29. U.S. Government Printing Office, Washington, DC.

Wagner, V., Bayer, P., Kübert, M. and Blum, P. (2012) Numerical sensitivity study of thermal response tests. *Renewable Energy*, **41**, 245–253.

Yavuzturk, C. (1999) *Modeling of vertical ground loop heat exchangers for ground source heat pump systems*. PhD thesis. Oklahoma State University.

Yavuzturk, C. and Spitler, J. (1999) A short time step response factor model for vertical ground loop heat exchangers. *ASHRAE Transactions*, **105**, 475–485.

7

Modeling Ground Heat Exchangers

Various models have been reported for heat transfer in borehole heat exchangers (BHEs), with three principal applications: design of BHEs (including sizing borehole depth and determining borehole numbers); analysis of in-situ ground thermal conductivity test data; and integration with building system models, that is, coupling the model with heating, ventilating, and air conditioning systems and building heat transfer models to determine performance. Changes in ground temperature and the borehole fluid often must be kept within acceptable limits over the life of the heat exchanger to ensure the heat pump operation. Modeling the heat flows in and outside the ground heat exchanger (GHE) can assist in system performance analysis by improving parameters related to the design of currently installed GHEs. In particular, a focus is placed on modeling of closed-loop systems in this chapter.

In some cases, modeling is performed to investigate the sustainability of the ground-source heat pump (GSHP) coupled to the GHEs. Furthermore, modeling of currently installed systems is needed to study the heat flow outside the heat exchanger field in cases such as the examination of potential environmental impact of low-temperature geothermal systems that could result from the heat flow in the ground surrounding heat exchangers. This information can ultimately provide guidance in regulating the installation of these systems. However, the details of such required regulations on the current system installation procedures are not included in this book.

Several simulation models for the heat transfer inside and outside the borehole are available. Based on how heat transfer from the borehole fluid to the surrounding ground is modelled and simulated, these modeling approaches can be divided into analytical and numerical. The models vary in the way the problem of heat conduction in the ground is solved, the way the interference between boreholes is treated and the way the methods are accelerated. Effects such as moisture migration and groundwater flow are also included in this chapter to explain their importance in modeling vertical ground heat exchangers (VGHEs) as well as to estimate the degree of complexity of the problem once they are included in the model.

7.1 General Aspects of Modeling

In solving the governing equations with both analytical and numerical approaches, parameters such as moisture migration and groundwater flow may need to be accounted for. In some cases, assumptions related to boundary conditions such as those at the

Geothermal Energy: Sustainable Heating and Cooling Using the Ground, First Edition.
Marc A. Rosen and Seama Koohi-Fayegh.
© 2017 John Wiley & Sons, Ltd. Published 2017 by John Wiley & Sons, Ltd.

ground surface need to be considered carefully as they could result in errors in the solution in some cases. In this section, such considerations are discussed.

7.1.1 Modeling Ground Surface Boundary Conditions

Modeling the ground surface involves consideration of several parameters such as solar radiation, cloud cover, surface albedo, ambient air temperature and relative humidity, rainfall, snow cover, wind speed, and evapotranspiration. Energy and moisture balances at the ground surface can be written that take into account the above parameters with the model of Tarnawski (1982), which involves complex processes. By accounting for the details a proper account of the renewable energy resource is provided.

The energy balance on the ground surface is written with the model of Tarnawski (1982) as:

$$q_n + q_{adv} + q_{sn} + q_h - q_{lo} - q_e = 0 \tag{7.1}$$

where q_n, q_{adv}, q_{sn}, q_h, q_{lo}, and q_e denote the heat flux by conduction from underground, advective energy (rain), net incoming shortwave radiation, convective heat transfer, net outgoing longwave radiation, and latent heat flux by evaporation, evapotranspiration, melting snow or sublimation, respectively. Of these fluxes, the radiation exchange is normally the most important while the convective heat transfer and heat flow by evaporation are typically of secondary importance.

However, due to the complexity of adding all the above heat fluxes to the numerical model, some studies assume the temperature variation at the ground surface to take the form of a sine wave or Fourier series (Jacovides *et al.* 1996; Mihalakakou and Lewis 1996; Salah El-Din 1999;Mihalakakou 2002). Others take the ground surface boundary to have a constant temperature equal to periodic air temperature, or isothermal at the ground temperature deep in the ground. Moreover, sometimes the problem is simplified further and an adiabatic boundary condition is assumed at the ground surface.

7.1.2 Moisture Migration in Soil

The moisture content of the soil surrounding a GHE can affect the rate and direction of heat flow in the ground surrounding heat exchangers. When considering the existence of moisture in the soil, the heat flux is described in terms of conduction, latent heat transport, and sensible heat flow.

The coupled phenomena of moisture migration and heat flow for soils have been explained by Philip and de Vries (1957). They combined the effects of liquid migration by capillary action and vapor diffusion due to temperature gradients into a dynamic mass balance. The heat flux is described in terms of conduction, latent heat transport, and sensible heat flow. Thermal energy and mass balances describe the coupled heat and moisture flows in soil. This complex problem requires a set of transient simultaneous partial differential equations to be solved, but many soil parameters are not readily available.

A modified and simpler model for the coupled heat and moisture flows in soils was developed involving non-isothermal mass transfer in an unsaturated soil in both liquid and vapor phases, driven by volumetric moisture content and a temperature gradient (Philip and de Vries 1957). This process can be described as follows:

$$\frac{\partial \theta}{\partial t} = \nabla \cdot (D_\theta \nabla \theta_l) + \nabla \cdot (D_T \nabla T) + \frac{\partial K_h}{\partial_y} \tag{7.2}$$

Here, θ, θ_l, D_θ, D_T, T, and K_h denote total volumetric moisture content, volumetric liquid content, isothermal moisture diffusivity, thermal moisture diffusivity, temperature, and unsaturated hydraulic conductivity, respectively. Heat transfer in soil can be represented by conduction, latent heat due to vapor pressure driven flux, and sensible heat carried by the liquid flux. The general governing differential equation for heat transfer in unsaturated soils can be expressed as follows:

$$C\frac{\partial T}{\partial t} = \nabla \cdot (k\nabla T) + L\varepsilon\frac{\partial}{\partial t}(\rho_l\theta) \tag{7.3}$$

where C, k, L, ε, and ρ_l denote volumetric heat capacity of soil, effective thermal conductivity of soil, latent heat of vaporization of water, phase conversion factor, and density of water, respectively. The simultaneous evaluation of Equation (7.2) and Equation (7.3) is challenging as they are coupled, highly non-linear partial differential equations. The equations can be solved using numerical approaches, however.

Research shows that the effects of moisture migration are normally not significant facors in the operation of a VGHE; these effects are usually more pronounced with a horizontal ground heat exchanger (HGHE). This is because natural variations of temperature and moisture near the ground surface and operation of the HGHE may create a potentially greater moisture movement. During the cooling season, migration of soil moisture away from the GHE may lead to a significant drop in soil thermal conductivity and consequently a significantly reduced heat transfer, which has a determinental effect on GHE performance. Therefore, although moisture migration effects can be neglected in early stages of design or conceptual development, not considering them in long-term operation of GSHP systems makes it difficult to assess the performance and failure potential of these systems (Leong and Tarnawski 2010).

Mei (1986) proposed a HGHE model to calculate thermal interaction between the circulating fluid and soil taking into account heat flow with or without moisture transfer in the soil. Piechowski (1999) also solved heat and moisture diffusion equations at the locations with the largest temperature and moisture gradients, that is, within a distance of 0.15 m from the pipe–soil interface. For the remaining soil region, the heat diffusion equation was applied only. Although the approach offers considerable reduction in simulation time, it is still time demanding, as small simulation time steps, on the order of minutes, are required. Therefore, this modeling approach is not suitable for simulating the long-term performance of large GSHP systems.

The freezing and thawing of moisture is modeled using the isothermal approach of Hromadka *et al.* (1981), which assumes that the energy interactions associated with the latent heat of fusion can be modeled for soil moisture as an isothermal process. In this approach, the temperature of a soil element during phase change is not allowed to change until the latent heat of all the soil moisture available for a phase change is removed or added to the control volume. Three cases can be considered for an isothermal freezing process: isothermal freezing with fixed temperature and moisture; isothermal freezing with moisture accumulation; and temperature change after freezing ends. This approach provides accurate simulation results while permitting a relatively large simulation time step (on order of hours) and spatial discretization.

The three-time-level scheme of Goodrich (1980), which provides the solution stability, accuracy and rapid convergence required for large simulation time steps such as hours or days, is used here with a simulation time step of 12 h.

Comprehensive processes for energy and moisture balances at the ground surface have been proposed by Tarnawski (1982), which account for many of the most relevant parameters: solar radiation; cloud cover; surface albedo; ambient air temperature; ambient relative humidity; rainfall; snow cover; wind speed; and evapotranspiration. As noted earlier, Tarnawski expresses the energy balance for the ground surface as shown in Equation (7.1). Rainfall is of little significance as an energy flux, but influences markedly the water content and other properties of soil (Leong *et al.*, 2011).

The simulation code Vertical Ground Heat Exchanger Analysis, Design and Simulation (VGHEADS) developed by Leong and Tarnawski (2010) determines the effects of simultaneous heat and moisture transfer on the performance of a solar-assisted GSHP with VGHEs. To quantify the effects of moisture transfer, two cases are compared with the model: independent conductive heat transfer; and simultaneous heat and moisture transfer. Neglecting moisture migration in soil is shown to cause a 2% difference in the annual heat rejection to the ground with respect to the case accounting for heat and moisture transfer.

7.1.3 Groundwater Movement

A further complication in the design of GSHP systems is the presence of groundwater. Due to the difficulties encountered both in modeling and computing the convective heat transfer and in understanding the actual groundwater flow in engineering practice, each of the methods presented in the previous sections is based on Fourier's law of heat conduction and neglect the effects of groundwater flow in carrying away heat. Where groundwater is present, flow will occur in response to hydraulic gradients, and the physical process affecting heat transfer in the ground is inherently a coupled one of heat diffusion (conduction) and heat advection by moving groundwater.

Underground water occurs in two zones: the unsaturated zone and the saturated zone. The term "groundwater" refers to the water in the saturated zone. The surface separating the saturated zone from the unsaturated zone is known as the "water table." At the water table, water in soil or rock pore spaces is at atmospheric pressure. In the saturated zone (below the water table), pores are fully saturated and water exists at pressures greater than atmospheric. In the unsaturated zone, pores are only partially saturated and the water exists under tension at pressures less than atmospheric. Groundwater is present nearly everywhere, but it is only when the local geology results in the formation of aquifers that significant flows of groundwater can be expected.

Aquifers are described as being either confined or unconfined. Confined aquifers are bounded between two or more layers of rock (or clay soils) of low permeability. Unconfined aquifers are bounded at their upper surface by the water table. In practice, the boreholes or ground-loop heat exchangers may partially penetrate unconfined aquifers and/or at greater depths penetrate confined aquifers. Knowledge of aquitard science with emphasis on aspects on groundwater resources use and management, investigations of aquitard integrity, and specific technical methodologies, categories of data collection, and synthesis are summarized by the AWWA Research Foundation (2006a, b).

In general, for material with high hydraulic conductivity and thus high discharge rates, steadily flowing groundwater is expected to be beneficial to the thermal performance of closed-loop GHEs. According to the conduction model, the required ground-loop heat exchanger lengths are significantly greater than the required lengths if the annual load were balanced so as to adequately dissipate the unbalanced annual loads. On the other hand, moderate groundwater advection is expected to make a notable difference in alleviating the possible heat buildup around the borehole over time. As a result, it is desirable to account for the groundwater flow in the heat transfer model to avoid over-sizing of GHEs. Therefore, it is essential to have tools that allow for the evaluation not only of technical aspects of GSHP systems but also the effects of groundwater flow on the system efficiency and, further, the temperature changes in the aquifer exerted by the energy extraction or injection rates.

Chiasson *et al.* (2000) analyze the effect of groundwater using a two-dimensional finite element scheme by discretizing a 4×4 borefield and including the interior of boreholes in the discretization. A simple but useful method of assessing the relative importance of heat conduction in the ground versus heat advection by moving groundwater is demonstrated through the use of the dimensionless Peclet number. Chiasson *et al.* utilize a finite element numerical groundwater flow and heat transfer model to simulate the effects of groundwater flow on a single closed-loop heat exchanger in various geologic materials. Their simulations show that the advection of heat by groundwater flow significantly enhances heat transfer in geologic materials with high hydraulic conductivity, such as sands, gravels, and rocks exhibiting fractures and solution channels.

Gehlin and Hellström (2003) investigate the groundwater effect on thermal response test of an infinite borehole using a two-dimensional finite difference method with regular square meshes. The borehole is represented by four squares.

Diao *et al.* (2004a) study the combined heat transfer by conduction and advection for vertical borehole heat exchangers with an analytical approach. Similar to Chiasson *et al.* (2000), they use a two-dimensional model of a borehole in an infinite porous medium with uniform water advection. They solve the model analytically by approximating the borehole by a line heat source and derive an explicit expression of the temperature response describing the correlation among various factors that impact this process. Compared with the conventional Kelvin's line-source model, which does not take into account water advection, this solution indicates that the impact of moderate groundwater flow on the heat transfer process may be significant. The actual magnitude of the impact, however, depends mainly on the flow rate, which can be characterized by a non-dimensional parameter. This explicit and concise expression can provide an appropriate basis for the qualitative and quantitative analysis of this impact for VGHEs in GSHP systems. The results of Diao *et al.* (2004a) show that the temperature rises at certain locations at the same distance from the source are independent of time.

Nam *et al.* (2008) use a numerical model that combines a heat transport model with groundwater flow to develop a heat exchanger model for an exact shape. The simulation code, FEFLOW, is used to calculate heat exchange rate between the GHE and the surrounding ground and to estimate the distribution of the subterranean temperature. The authors validate their numerical simulation technique by comparing the results with available experimental data, and find good agreement between the two.

MT3DMS is a widely used program for the simulation of solute transport in porous media. Owing to the mathematical similarities between the governing equations for solute transport and heat transport, this program appears also applicable to the simulation of thermal transport phenomena in saturated aquifers. Hecht-Mendez *et al.*

(2010) evaluate simulations of a single borehole GSHP system for three scenarios: a pure conduction situation; an intermediate case; and a convection-dominated case. Two evaluation approaches are employed. First, MT3DMS heat transport results are compared with analytical solutions. Secondly, finite difference simulations by MT3DMS are compared with those using the finite element code FEFLOW and the finite difference code SEAWAT. The results suggest that MT3DMS can successfully simulate GSHP systems, and likely other systems with similar temperature ranges and gradients in saturated porous media.

7.2   Analytical Models

In this section, various analytical models that are used to calculate heat transfer characteristics of GHEs are examined, ranging from primarily one-dimensional ones to two- and three-dimensional models which have been devised in recent years. The heat transfer modeling in GHEs via analytical methods is complicated since their study involves transient effects in a time range of months or even years. Because of the complexities of this problem and its long time scale, heat transfer in GHEs is usually analyzed in two separated regions (Figure 7.1): the region inside the borehole containing the U-tubes and the grout (Zeng *et al.* 2003a), and the ground surrounding the borehole. The transient borehole wall temperature is important for engineering applications and system simulation. It can be determined by modeling the region outside the borehole by various methods such as cylindrical heat source theory (Carslaw and Jaeger 1946). Based on the borehole wall temperature, the fluid inlet and outlet temperatures can be evaluated by a heat transfer analysis inside the borehole. In other words, the regions inside and outside the borehole are coupled by the borehole wall temperature. The heat pump model utilizes the fluid inlet and outlet temperatures for the GHE, and accordingly the dynamic simulation and optimization design for a GSHP system can be implemented. This is the basic idea behind the development of the two-region VGHE model. Some models combine ground heat conduction outside the borehole

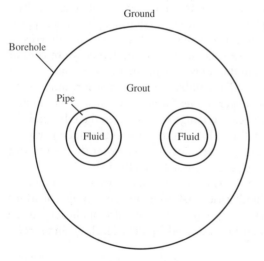

Figure 7.1 Cross section of vertical ground heat exchanger.

with borehole heat exchanger models to predict heat extraction/injection rates from/to the ground (Jun *et al.* 2009). Yang *et al.* (2010) present a detailed summary of the most typical simulation models of the VGHEs, including heat transfer processes outside and inside boreholes.

7.2.1 Heat Transfer Inside the Borehole

While most models investigate thermal characteristics of ground surrounding the borehole, a few models of varying complexity in how they deal with the complicated geometry inside boreholes have been established to describe heat transfer within VGHEs. The thermal analysis in the borehole seeks to define the inlet and outlet temperatures of the borehole fluid according to borehole wall temperature, its heat flow and the thermal resistance inside the borehole. The latter quantity is determined by thermal properties of the grouting material, the arrangement of flow channels and the convective heat transfer in the tubes. If the thermal resistance between the borehole wall and borehole fluid is determined, the GHE fluid temperature can be calculated. In the absence of natural convection, moisture flow and freezing, the borehole thermal resistance can be calculated assuming steady-state heat conduction in the region between the borehole fluids and a cylinder around the borehole.

According to Jun *et al.* (2009), a steady-state heat transfer assumption is made when the running time is greater than the critical time, that is Fo > 5, where Fo is the Fourier number, and the impact of thermal capacity of objects inside the borehole can be neglected. Such a simplification has been proven reasonably accurate and convenient for most engineering practices, except for analyses dealing with dynamic responses within a few hours (Yavuzturk 1999).

7.2.1.1 One-Dimensional Model
In this model, the axial heat flows in the grout and pipe walls are considered negligible as the borehole dimensional scale is small compared with the infinite extent of the ground beyond the borehole (Bose *et al.* 1985). As a consequence of the U-tube structure, the heat conduction in the cross section is clearly two-dimensional, which is somewhat complicated to solve. Therefore, simplified models conceiving the U-tube as a single pipe have been recommended and heat transfer in the borehole is approximated as a steady-state one-dimensional process. The thermal resistance inside the borehole R_b can be defined as the sum of the thermal resistance of the fluid convection, and the thermal resistances of conduction in the pipe R_p and in the grout R_g:

$$R_b = R_p + R_g \tag{7.4}$$

The thermal resistance of the fluid convection and conduction in the pipe is defined as:

$$R_p = \frac{1}{2\pi r_i h_i} + \frac{\ln(r_p/r_i)}{2\pi k_p} \tag{7.5}$$

where h_i is determined by the Dittus–Boelter correlation:

$$h_i = \frac{0.023 \, \mathrm{Re}^{0.8} \, \mathrm{Pr}^n k_f}{d_i} \tag{7.6}$$

The first term on the right-hand side of Equation (7.5) accounts for the resistance due to fluid convection and the second term accounts for conduction in the pipe. The thermal resistance of the grout can be computed by the equivalent diameter method or the shape factor method. In the first, the two legs of the U-tube are considered as one concentric cylindrical heat source/sink, also referred to as an "equivalent pipe" having identical temperatures inside the borehole, which leads to the following simple expression for the grout thermal resistance:

$$R_g = \frac{1}{2\pi k_b} \ln\left(\frac{d_b}{d_e}\right) \tag{7.7}$$

where d_b is the borehole diameter, d_e is the equivalent diameter, and k_b is the thermal conductivity of the grout.

Claesson and Dunand (1983) give the equivalent diameter as:

$$d_e = \sqrt{2}d_p \tag{7.8}$$

while Gu and O'Neal (1998) suggest the following expression:

$$d_e = \sqrt{2d_p L_s} \quad (d_p < L_s < r_b) \tag{7.9}$$

Note that when the equivalent diameter method is used for computing the thermal resistance inside a borehole, the thermal resistance of fluid convection and conduction in the pipe should remain constant.

In reality, the fluid circulating through different legs of the U-tube exchanges heat with the surrounding ground and is of varying temperature along the tube. Therefore, thermal interference, that is, thermal "short-circuiting," among U-tube legs, which degrades the effective heat transfer in the GHEs, is inevitable. This oversimplified one-dimensional model is not capable of evaluating this impact or analyzing dynamic responses within a few hours.

Paul (1996) expresses the grout resistance using the concept of the shape factor of conduction as follows:

$$R_g = \left[k_b \beta_0 \left(\frac{r_b}{r_p}\right)^{\beta_1}\right]^{-1} \tag{7.10}$$

where β_0 and β_1 are the shape factors of the grout resistance whose values depend on the relative location of U-tube pipes in the borehole. These factors are obtained by curve fitting of measured effective borehole resistances determined in laboratory measurements. In this approach only a limited number of influencing factors are considered, and all pipes are assumed to be of identical temperature as a precondition.

7.2.1.2 Two-Dimensional Model

Due to the axial convective heat transport and the transverse heat transfer to the ground, the temperature of the fluid varies along a U-tube. In particular, when the flow rate is low, there is a bigger temperature difference between the upward and downward channels which may result in heat exchange between the two channels and a reduced efficiency of the GHE. Hellström (1991) takes into account the thermal resistances among pipes in the cross section perpendicular to the borehole axis and obtains a two-dimensional analytical solution to the heat transfer problem inside the borehole.

This model is superior to the oversimplified one-dimensional models in presenting quantitative expressions of the thermal resistance in the cross section and for describing the impact of the U-tube placement on heat conduction.

In the two-dimensional model, the temperature of the fluid in the U-tube is defined by superposing two separate temperature responses caused by the heat fluxes per unit length, q_1' and q_2', from the two pipes of the U-tube. The fluid temperatures in the U-tubes (T_{f1} and T_{f2}) can be determined as follows:

$$T_{f1} - T_b = R_{11}q_1' + R_{12}q_2'$$
$$T_{f2} - T_b = R_{12}q_1' + R_{22}q_2' \tag{7.11}$$

where t_b is the temperature on the borehole wall, R_{11} and R_{22} are the thermal resistances between the borehole fluid in each pipe and the borehole wall, and R_{12} is the resistance between the two pipes. Note that the temperature on the borehole wall (T_b) is assumed uniform along the borehole depth and is taken to be a reference of the temperature excess. A linear transformation of Equation (7.11) leads to:

$$q_1' = \frac{T_{f1} - T_b}{R_1^\Delta} + \frac{T_{f1} - T_{f2}}{R_{12}^\Delta}$$
$$q_2' = \frac{T_{f2} - T_b}{R_2^\Delta} + \frac{T_{f2} - T_{f1}}{R_{12}^\Delta} \tag{7.12}$$

where

$$R_1^\Delta = \frac{R_{11}R_{22} - R_{12}^2}{R_{22} - R_{12}}, \quad R_2^\Delta = \frac{R_{11}R_{22} - R_{12}^2}{R_{11} - R_{12}}, \quad \text{and} \quad R_{12}^\Delta = \frac{R_{11}R_{22} - R_{12}^2}{R_{12}} \tag{7.13}$$

Note that there is no distinction between the entering and exiting pipes since this model does not take into account the heat transfer of the borehole fluid in the axial direction. Eskilson (1987) determines thermal resistance between the fluid and borehole wall as:

$$R_{b2} = \frac{R_{11} + R_{12}}{2} \tag{7.14}$$

By assuming identical temperatures and heat fluxes of the pipes in the borehole:

$$T_{f1} = T_{f2} = T_f \text{ and } q_1' = q_2' = q_l'/2 \tag{7.15}$$

the borehole resistance is derived for symmetrically placed double U-tubes as:

$$R_b = \frac{1}{2\pi k_b}\left[\ln\left(\frac{r_b}{r_p}\right) - \frac{3}{4} + \left(\frac{D}{r_b}\right)^2 - \frac{1}{4}\ln\left(1 - \frac{D^8}{r_b^8}\right)\right.$$
$$\left. - \frac{1}{2}\ln\left(\frac{\sqrt{2}D}{r_p}\right) - \frac{1}{4}\ln\left(\frac{2D}{r_p}\right)\right] + \frac{R_p}{4} \tag{7.16}$$

The two-dimensional model presents quantitative expressions of the thermal resistance in the cross section, and permits descriptions of the impact of the U-tube placement in the borehole on conduction. However, the assumption of identical temperature for all pipes prevents this model from revealing the impact of the thermal interference on GHE performance.

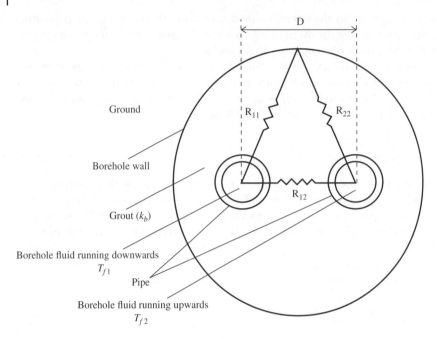

Figure 7.2 Thermal resistances in the borehole.

Zeng *et al.* (2003b) discuss the impact of thermal interference between the U-tube pipes and show that the thermal "short-circuit" phenomena may reduce heat transfer between the heat carrier fluid and ground, and deteriorate the performance of GHEs.

7.2.1.3 Quasi-Three-Dimensional Model

A quasi-three-dimensional model was proposed by Zeng *et al.* (2003a, b) taking into account the fluid axial convective heat transfer and thermal "short-circuiting" among U-tube legs. Being minor in order, conductive heat flow in the grout and ground in the axial direction is still neglected to keep the model concise and analytically manageable. Energy balance equations for up-flow and down-flow of the borehole fluid can be written, respectively, as:

$$-\dot{m}c_p\frac{dT_{f1}}{dz} = \frac{T_{f1} - T_b}{R_1^\Delta} + \frac{T_{f1} - T_{f2}}{R_{12}^\Delta}$$
$$\dot{m}c_p\frac{dT_{f2}}{dz} = \frac{T_{f2} - T_b}{R_2^\Delta} + \frac{T_{f2} - T_{f1}}{R_{12}^\Delta} \qquad (0 \leq z \leq H) \qquad (7.17)$$

where T_{f1}, T_{f2}, and T_b are the temperatures of the fluid running downwards, of the fluid running upwards, and of the borehole wall, respectively, z is the direction along the tube, and R_1^Δ, R_{12}^Δ, and R_2^Δ are the thermal resistances defined in Equation (7.13).

Here, R_{11} and R_{22} are the thermal resistances between the borehole fluid and the borehole wall, and R_{12} is the resistance between the pipes (Figure 7.2). In most engineering applications, the configuration of the U-tube in the borehole may be assumed symmetric and, when it is assumed that $R_{22} = R_{11}$, the following relations can be derived:

$$R_1^\Delta = R_2^\Delta = R_{11} + R_{12}, \quad R_{12}^\Delta = \frac{R_{11}^2 - R_{12}^2}{R_{12}} \tag{7.18}$$

The steady-state conduction problem in the borehole cross-section was analyzed by Hellström (1991) and Claesson and Hellström (2011) with the line source and Multipole approximations. The line-source assumption results in the following solution:

$$R_{11} = \frac{1}{2\pi k_b} \left[\ln\left(\frac{r_b}{r_p}\right) + \frac{k_b - k}{k_b + k} \cdot \ln\left(\frac{r_b^2}{r_b^2 - D^2}\right) \right] + R_p$$

$$R_{12} = \frac{1}{2\pi k_b} \left[\ln\left(\frac{r_b}{2D}\right) + \frac{k_b - k}{k_b + k} \cdot \ln\left(\frac{r_b^2}{r_b^2 - D^2}\right) \right] \tag{7.19}$$

where r_b, r_p, k_b, k, D, and R_p are the radius of the boreholes, radius of the pipe, the grout thermal conductivity, the ground thermal conductivity, the distance between the pipes in the borehole, and the thermal resistance of conduction in the pipe, respectively (Figure 7.2).

The following boundary conditions are applied to the energy equations [Equation (7.17)]:

$$z = 0, \quad T_{f1} = T_f'$$
$$z = H, \quad T_{f1} = T_{f2} \tag{7.20}$$

where T_f' is the temperature of the fluid entering the U-tube. Using a Laplace transformation, the general solution of Equation (7.17) is obtained, which is complicated in form and can be found elsewhere (Zeng *et al.* 2003b). At the instance of symmetric placement of the U-tube inside the borehole, the temperature profiles in the two pipes are reduced as follows:

$$\Theta_1(Z) = \cosh(\beta Z) - \frac{1}{\sqrt{1 - P^2}} \left[1 - P \frac{\cosh(\beta) - \sqrt{\frac{1 - P}{1 + P}} \sinh(\beta Z)}{\cosh(\beta) + \sqrt{\frac{1 - P}{1 + P}} \sinh(\beta)} \right] \sinh(\beta Z)$$

$$\Theta_2(Z) = \frac{\cosh(\beta) - \sqrt{\frac{1 - P}{1 + P}} \sinh(\beta)}{\cosh(\beta) + \sqrt{\frac{1 - P}{1 + P}} \sinh(\beta)} \cosh(\beta Z) \tag{7.21}$$

$$+ \frac{1}{\sqrt{1 - P^2}} \left[\frac{\cosh(\beta) - \sqrt{\frac{1 - P}{1 + P}} \sinh(\beta)}{\cosh(\beta) + \sqrt{\frac{1 - P}{1 + P}} \sinh(\beta)} - P \right] \sinh(\beta Z)$$

where the dimensionless parameters are defined as:

$$\Theta = \frac{T_f(z) - T_b}{T'_f - T_b}, \quad Z = \frac{z}{H}, \quad P = \frac{R_{12}}{R_{11}}$$

$$\beta = \frac{H}{\dot{m}c_p \sqrt{(R_{11} + R_{12})(R_{11} - R_{12})}}$$

(7.22)

Zeng *et al.* (2003a) illustrate the temperature profiles in the pipes for the symmetric placement of the U-tube inside the borehole. They present a parameter called the heat transfer efficiency of the borehole as:

$$\varepsilon = \frac{T'_f - T''_f}{T''_f - T_b}$$

(7.23)

which deals with the heat exchanger inside the borehole only, and is independent of time. Here, T''_f is the temperature of the fluid exiting the U-tube. They obtain analytical expressions for the efficiency of a vertical geothermal heat exchanger for the two-dimensional and quasi-three-dimensional cases and show the fluid temperature differences along the borehole axial direction.

Taking the heat transfer efficiency of the borehole (ε) into account in the temperature profile derived from Equation (7.17), Diao *et al.* (2004b) derive the thermal resistance between the fluid inside the U-tube and the borehole wall as:

$$R_{b3} = \frac{H}{\dot{m}c_p}\left(\frac{1}{\varepsilon} - \frac{1}{2}\right)$$

(7.24)

where R_{b3} is the borehole thermal resistance for the three-dimensional analysis. They concede that the relative error in the borehole resistance between the two-dimensional and quasi-three-dimensional models is a function of the single dimensionless parameter β only. The value of β is usually less than 0.6, resulting in an error of less than 11%.

Zeng *et al.* (2003b) focus on quasi-three-dimensional heat transfer inside a borehole with double U-tubes, and determine analytical expressions of the thermal resistance of single and double U-tube boreholes for all possible circuit layouts. Comparisons of the performances of single and double U-tube boreholes show that double U-tube boreholes are superior to single U-tube boreholes, with reductions in borehole resistance of 30–90%. Also, superior performance is observed in double U-tubes in parallel compared with those in series. Zeng *et al.* (2003b) also examine a relatively wide range of factors, including geometrical parameters (borehole and pipe sizes and pipe placement in the borehole), physical parameters (thermal conductivity of the materials, flow rates, and fluid properties) and the flow circuit configuration.

Quasi-three-dimensional models reveal drawbacks of two-dimensional models and are thus preferred for design and analysis of GHEs, as they provide more accurate information for performance simulation and analysis and design. Diao *et al.* (2004b) discuss and summarize the improvements in modeling of VGHE from the aspect of heat transfer analysis inside the borehole. Analytical methods for modeling the heat transfer inside the borehole are summarized in Table 7.1.

7.2.2 Heat Transfer Outside the Borehole

Several simulation models for the heat transfer outside the borehole are available, most of which are based on analytical and/or numerical methods. The models vary in the way

Table 7.1 Comparison of various methods for heat transfer analysis inside a borehole.

Heat transfer characteristic	One-dimensional (Equivalent diameter)	One-dimensional (Shape factor)	Two-dimensional	Quasi-three-dimensional
U-tube placement in the borehole	N	Y	Y	Y
Quantitative expressions of the thermal resistance in the cross section	N	N	Y	Y
Thermal interference	N	N	N	Y
Distinction between the entering and exiting pipes	N	N	N	Y
Axial convection by fluid flow	N	N	N	Y
Axial conduction in grout	N	N	N	N

Y, yes; N, no.

the problem of heat conduction in the ground is solved, the way the interference between boreholes is treated and the way the methods are accelerated. In the analysis of GHE heat transfer, some complicating factors, such as ground stratification, ground temperature variation with depth, and groundwater movement (Chiasson *et al*. 2000) usually prove to be of minor importance and are analyzed separately. As a basic problem, the following assumptions are commonly made:

- The ground is homogeneous in its thermal properties and initial temperature.
- Moisture migration is negligible.
- Thermal contact resistance is negligible between the pipe and the grout, the grout and the borehole wall, and the borehole wall and the ground adjacent to the borehole wall.
- The effect of ground surface is negligible.

Unlike the area inside the borehole, heat conduction outside the borehole exhibits transient behavior. The thermal response due to a step-change in the specific heat injection rate $\dot{q}$ given per unit length of the borehole associated with a temperature evolution $(T_b - T_0)$ results in a time-dependent ground thermal resistance R_g.

The heat conduction in the radial direction is dominant when there is no groundwater flow and the effect of the ground surface can be neglected for the initial 5–10 years (depending on the borehole depth). Therefore, the heat transfer is usually modeled with a one-dimensional line-source (Eskilson 1987) or cylindrical-source theory (Carslaw and Jaeger 1946), which are discussed in Section 6.2.2. In both models, the borehole depth is considered infinite and axial heat flow along the borehole depth is assumed negligible.

In terms of accuracy, such analytical models are equivalent to numerical approaches in estimating the temperature of the ground surrounding boreholes. Since the borehole

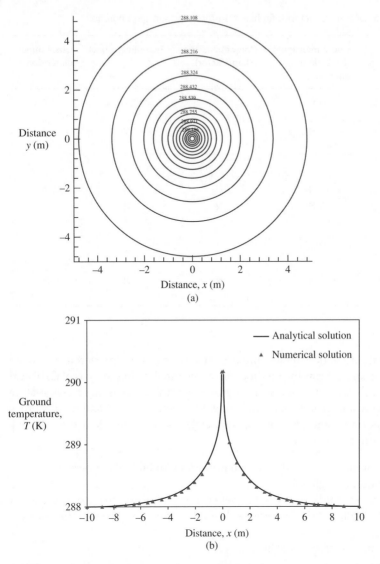

Figure 7.3 Ground temperatures (K) for a single borehole at time $t = 6$ months. (a) Temperature contours (K) of the analytical solution. (b) Comparison of the analytical and numerical solutions at $y = 0$.

is assumed to have an infinite length in these analytical methods, it is adequate for the numerical method to evaluate the temperature rise in the ground surrounding a borehole in a two-dimensional plane perpendicular to the axial direction of the borehole (Figure 7.3a). The results of a finite volume method used to evaluate the temperature of the ground surrounding a borehole is compared with those of the line-source theory in Figure 7.3b. The temperature rise in the ground is evaluated for 6 months of system operation with 3.1 W of heat flow rate per unit length of borehole. Since the simulations in this section are performed to compare the numerical solution with the results of the line-source solution, a somewhat smaller heat flow rate per unit length is chosen here to reduce the size of the solution domain and computation time. Typical values of heat flow rate per unit length for borehole heat exchangers are within $30-60\ \text{W/m}$ and require a much larger computational domain. As can be seen, the temperature reaches a maximum at the borehole wall and decreases with the distance from the borehole wall. The numerical values for the temperature around the borehole can be seen to agree with the analytical ones. The temperature of the ground just beside the borehole wall is calculated as 290.2 K for the numerical solution. This temperature is calculated as 289.6 K for the analytical solution. Similarly, the affected region in the ground, that is, where the temperature excess in the ground exceeds 0.1 K, after 6 months of heat injection into the ground is about 5 m for both solution methods. The area of the affected region, however, depends highly on the heat flow rate per unit length of the borehole; for typical borehole heat flow rates, a larger affected area is expected.

When time tends to infinity, the temperature rise of Kelvin's theory for an infinite line source tends to infinity, making the infinite model weak for describing the heat transfer mechanism at long time steps. On the other hand, the temperature from the finite line-source model approaches the steady state corresponding to the actual heat transfer mechanism. Therefore, they can only be used for short operation durations of GSHP systems. To take into account axial temperature changes for boreholes with finite lengths and in long durations, a number of approaches for ground-loop heat exchangers have been devised that combine numerical and analytical methods.

Eskilson's approach to the problem of determining the temperature distribution around a borehole is based on a hybrid model combining analytical and numerical solution techniques. Eskilson (1987) applies a numerical finite difference method to the transient radial-axial heat conduction equation for a single borehole:

$$\frac{\partial^2 T}{\partial r^2} + \frac{1}{r}\frac{\partial T}{\partial r} + \frac{\partial^2 T}{\partial z^2} = \frac{1}{a}\frac{\partial T}{\partial t} \tag{7.25}$$

assuming no temperature change on the ground surface (by superposing an identical mirror borehole above the ground surface with negative strength). Note that, compared with Equation (6.2), the third term on the left-hand side of Equation (7.25) accounts for the axial heat flow along the borehole depth. Assuming the ground to be homogeneous with constant initial and boundary temperatures, the boundary conditions are:

$$T(r, 0, \tau) = T_0$$

$$T(r, z, 0) = T_0$$

$$q'(t) = \frac{1}{H} \int_D^{D+H} 2\pi r k \frac{\partial T}{\partial r}\Big|_{r=r_b} dz \tag{7.26}$$

where H is the active borehole length and D is the uppermost part of the borehole. The thermal capacitance of the individual borehole elements such as the tube wall and the grout are neglected. The temperature fields from a single borehole are superposed in space to obtain the response from the borefield. The temperature response of the borefield is converted to a set of non-dimensional temperature response factors, called g-functions. The g-function allows the calculation of the temperature change at the borehole wall in response to a step heat input for a time step. Once the response of the borefield to a single step heat pulse is represented with a g-function, the response to any arbitrary heat rejection/extraction function can be determined by devolving the heat rejection/extraction into a series of step functions, and superposing the response to each step function. Therefore, the temperature distribution at the wall of a single borehole with finite length to a unit step heat pulse is defined as:

$$T_b - T_0 = -\frac{q'}{2\pi k}g(t/t_s, r_b/H) \tag{7.27}$$

where $t_s = H^2/9\alpha$ is the steady-state time and the g-function is the non-dimensional temperature distribution at the borehole wall, which is computed numerically. The g-function curves are developed based on selected borefield configurations.

For the temperature responses of multiple boreholes, using a superposition method in space to determine the overall temperature response of the GHE, g-functions of the GHEs with different configurations (i.e., any heat rejection/extraction at any time) have to be pre-computed and stored in the program as a large database with one of the parameters fixed. Therefore, an interpolation function is applied when using the database causing some computing errors. The model is intended to provide the response of the ground to heat rejection/extraction over longer periods of time (up to 25 years). Since the numerical model that provides the g-functions does not account for the local borehole geometry, it cannot accurately provide shorter term responses.

Modifying Kelvin's line-source model, Zeng *et al.* (2002, 2003a) and Diao *et al.* (2004b) present an analytical solution to the transient finite line-source problem considering the effects of the finite borehole length and the ground surface as a boundary. Their study is based on the following assumptions:

- The temperature of the ground surface remains constant and equal to its initial value (T_0) over the time period concerned.
- The heating rate per length of the source (q') is constant.

With these assumptions, the non-dimensional solution of the temperature excess is:

$$\theta(\overline{R}, Z, \text{Fo}) = \frac{q'}{4k\pi}\int_0^1 \left\{ \frac{\text{erfc}\left(\frac{\sqrt{\overline{R}^2 + (Z-\overline{H})^2}}{2\sqrt{\text{Fo}}}\right)}{\sqrt{\overline{R}^2 + (Z-\overline{H})^2}} - \frac{\text{erfc}\left(\frac{\sqrt{\overline{R}^2 + (Z+\overline{H})^2}}{2\sqrt{\text{Fo}}}\right)}{\sqrt{\overline{R}^2 + (Z+\overline{H})^2}} \right\} d\overline{H} \tag{7.28}$$

where the dimensionless parameters are defined as:

$$\theta = T - T_0, \quad Z = \frac{z}{H}, \quad \overline{H} = \frac{h_z}{H}, \quad \overline{R} = \frac{r}{H}, \quad \text{Fo} = \frac{\alpha t}{H^2} \tag{7.29}$$

and the integral can be computed numerically. A comparison of the analytical results and the numerical data from Eskilson's solution show satisfactory agreement, especially when $\alpha t/r_b^2 \geq 5$. With respect to long durations, the explicit solution of a finite line-source model better describes the temperature responses of the borehole for long time steps.

Furthermore, in both Eskilson's model and the finite line-source model, the radial dimension of the borehole and, therefore, the thermal capacity of the borehole, including the U-tubes, the borehole fluid, and the grout, are neglected. Eskilson estimates that the results for temperature responses on the borehole wall due to this assumption are only valid for a time greater than $5r_b^2/\alpha$, where the terms are as defined earlier.

It can be seen from Equation (7.28) that the temperature of the medium varies with time, radial distance from the borehole, and borehole depth. A representative temperature for the borehole wall is often chosen which represents the mean borehole wall temperature along the borehole depth and is used in the heat transfer analysis inside the borehole. To choose a representative temperature for the borehole wall ($r = r_b$) along the borehole depth, one can either choose the temperature at the middle of the borehole depth ($z = 0.5H$) or the integral mean temperature along the borehole depth, which may be determined by numerical integration of Equation (7.28). The difference between the two is analyzed and found to be insignificant (Zeng *et al.* 2002).

Yang *et al.* (2009) propose and develop an updated two-region vertical U-tube GHE analytical model. It divides the heat transfer region of the GHE into two parts at the boundary of the borehole wall, and the two regions are coupled by the temperature of the borehole wall. They use both steady and transient heat transfer methods to analyze the heat transfer process inside and outside the borehole, respectively. To model the region outside the borehole, they use cylindrical-source theory and for the region inside the borehole a quasi-three-dimensional model. Both models are coupled by the transient temperature of the borehole wall. An experimental validation of the model indicates that the calculated fluid outlet temperatures of the GHE agree well with the corresponding test data and the relative error is less than 6%.

Cui *et al.* (2006) establish a transient three-dimensional heat conduction model to describe the temperature response in the ground caused by a single inclined line source. Heat transfer in the GHEs with multiple boreholes is then studied by superposition of the temperature excesses that result from individual boreholes. The thermal interference between inclined boreholes is compared with that between vertical ones. The analyses provide a basic and useful tool for the design and thermal simulation of GHEs with inclined boreholes.

To deal with loads varying with time, the method of load aggregation can be employed (Bernier *et al.* 2004). The load profile is divided into various constant load steps starting at particular time instants. The overall performance is the summation of effects from each load step. Bernier *et al.* (2004) suggest a multiple load aggregation algorithm to calculate the performance of a single borehole at variable load based on the cylindrical-source model.

Hellström (1991) proposes a simulation model for vertical ground heat stores, which are densely packed ground-loop heat exchangers used for seasonal thermal energy storage. This type of system may or may not incorporate heat pumps to heat buildings. This model divides the medium with multiple boreholes into two regions: "local" which is the region surrounding a single borehole and "global" which is the farfield beyond the bulk volume of the multiple boreholes. He calculates the store performance based on a steady flux solution and solutions for the local and global regions. The numerical model used for the global region is a two-dimensional explicit finite difference scheme in the radial–axial coordinate system.

Kavanaugh (1995) proposes an equation for calculating the required total borehole length by including various terms into the steady-state heat transfer equation to account for load cycle effect, and thermal interference from adjacent boreholes and within tubes inside a borehole. These methods estimate the performance of the entire borefield.

Bandyopadhyay *et al.* (2008) developed a three-dimensional model to simulate BHEs. They obtain a semi-analytical solution for the U-tube geometry in grouted boreholes using a Laplace transform and subsequent numerical inversion of the Laplace domain solution. The solution results are found to agree with the results from the finite element method.

7.2.2.1 Heat Flow Rate Variation along the Borehole

The model presented by Zeng *et al.* (2002, 2003a) leads to an analytical relation for the temperature excess in the ground assuming a constant heat flow rate on the borehole wall (here, the line source). Modifying this model slightly to account for the variation of heat flow rate along the line source $q'(\overline{H})$, the temperature profile in the ground surrounding the boreholes is calculated as:

$$\theta(\overline{R}, Z, \text{Fo}) = \frac{1}{4k\pi} \int_0^1 q'(\overline{H})$$

$$\times \left[\frac{\text{erfc}\left(\frac{\sqrt{\overline{R}^2 + (Z - \overline{H})^2}}{2\sqrt{\text{Fo}}} \right)}{\sqrt{\overline{R}^2 + (Z - \overline{H})^2}} - \frac{\text{erfc}\left(\frac{\sqrt{\overline{R}^2 + (Z + \overline{H})^2}}{2\sqrt{\text{Fo}}} \right)}{\sqrt{\overline{R}^2 + (Z + \overline{H})^2}} \right] d\overline{H} \quad (7.30)$$

where the dimensionless parameters are defined as:

$$\theta = T - T_0, \quad Z = \frac{z}{H}, \quad \overline{H} = \frac{h_z}{H}, \quad \overline{R} = \frac{r}{H}, \quad \text{Fo} = \frac{\alpha t}{H^2} \quad (7.31)$$

Also, $q'(\overline{H})$ denotes the heating strength per unit length, t the time from the start of operation, α the thermal diffusivity of ground, z the axis along the borehole length, r the radial axis, H the borehole heating length, h_z the depth at which borehole heating starts, T_0 ground initial temperature, and T the temperature of the ground.

This solution [Equation (7.30)] is used as the basis for more complicated cases such as those with a system of n boreholes or with time-varying heat transfer rates.

7.2.2.2 Modeling Multiple Boreholes

Since the conduction equation is linear, the temperature response in the ground surrounding multiple boreholes can be calculated by superposing the temperature rise in the ground caused by individual boreholes. It is shown in Figure 7.4 that ground temperatures evaluated by superposing temperatures based on two analytical methods for two neighboring boreholes agree well with temperature evaluations based on a finite volume numerical method used in a two-dimensional plane perpendicular to the axial direction of the boreholes. Therefore, the effect of the temperature rise due to one borehole on the thermal performance of other boreholes can be neglected. Figure 7.5 shows the temperature contours in the ground surrounding two boreholes at $t = 1$ month and 6 months.

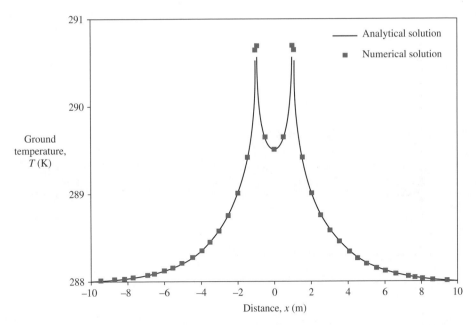

Figure 7.4 Comparison of the ground temperature (K) around multiple boreholes evaluated by analytical and numerical solutions at $y = 0$ m at $t = 6$ months.

Note that the affected region around the two boreholes (temperature excess of more than 0.1 K) grows with time (from 3 m at $t = 1$ month to 6 m at $t = 6$ months). Furthermore, the temperature of the ground immediately outside the borehole wall increases from 288 K at $t = 0$ to 289.9 K after 1 month and to 290.7 K after 6 months of heat injection into the ground.

The example shown in Figure 7.4 and Figure 7.5 is used here only to show the application of superposition of the temperature rise associated with operation of two single boreholes. For simplicity, the simple solution for line-source theory, described in Section 6.2.2.1, is superposed in this example. However, more accurate solutions accounting for axial temperature gradients in addition to radial temperature gradients [Equation (7.30)] can also be superposed to model the temperature response surrounding multiple boreholes. For the general case of a borehole system of n boreholes, the temperature response in the ground surrounding the borehole system can be calculated by superposing the temperature response evaluated by each borehole from Equation (7.30):

$$\theta(R, Z, \mathrm{Fo}) = \sum_{1}^{n} \int_{0}^{1} \frac{q_i'(\overline{H})}{4k\pi}$$
$$\left[\frac{\mathrm{erfc}\left(\dfrac{\sqrt{\overline{R}_i^2 + (Z_i - \overline{H})^2}}{2\sqrt{\mathrm{Fo}}} \right)}{\sqrt{\overline{R}_i^2 + (Z_i - \overline{H})^2}} - \frac{\mathrm{erfc}\left(\dfrac{\sqrt{\overline{R}_i^2 + (Z_i + \overline{H})^2}}{2\sqrt{\mathrm{Fo}}} \right)}{\sqrt{\overline{R}_i^2 + (Z_i + \overline{H})^2}} \right] d\overline{H} \qquad (7.32)$$

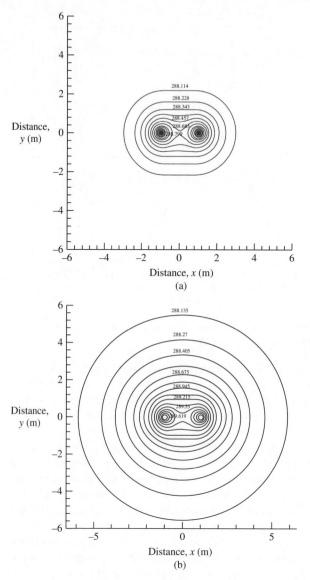

Figure 7.5 Ground temperature contours (K) of the analytical solution for multiple boreholes at time (a) $t = 1$ month and (b) $t = 6$ months.

where q'_i is the heat flow rate per unit length of borehole i (Figure 7.6) and

$$\bar{R}_i = \frac{\sqrt{(x - l_i)^2 + (y - w_i)^2}}{H} \tag{7.33}$$

where l_i and w_i refer to the position of borehole i in x and y coordinates, respectively (Figure 7.6). For the two boreholes that are shown in Figure 7.7, Equation (7.32) can be simplified to:

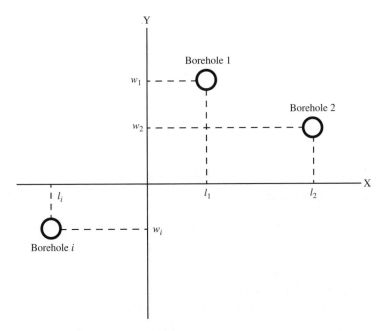

Figure 7.6 Definition of geometric parameters l_i and w_i in Equation (7.33).

$$\theta(\overline{R}, Z, \text{Fo}) = \theta_1(\overline{R}_1, Z, \text{Fo}) + \theta_2(\overline{R}_2, Z, \text{Fo}) =$$

$$\int_0^1 \frac{q'(\overline{H})}{4k\pi} \left[\frac{\text{erfc}\left(\dfrac{\sqrt{\overline{R}_1^2 + (Z - \overline{H})^2}}{2\sqrt{\text{Fo}}} \right)}{\sqrt{\overline{R}_1^2 + (Z - \overline{H})^2}} - \frac{\text{erfc}\left(\dfrac{\sqrt{\overline{R}_1^2 + (Z + \overline{H})^2}}{2\sqrt{\text{Fo}}} \right)}{\sqrt{\overline{R}_1^2 + (Z + \overline{H})^2}} \right] d\overline{H}$$

$$+ \int_0^1 \frac{q'(\overline{H})}{4k\pi} \left[\frac{\text{erfc}\left(\dfrac{\sqrt{\overline{R}_2^2 + (Z - \overline{H})^2}}{2\sqrt{\text{Fo}}} \right)}{\sqrt{\overline{R}_2^2 + (Z - \overline{H})^2}} - \frac{\text{erfc}\left(\dfrac{\sqrt{\overline{R}_2^2 + (Z + \overline{H})^2}}{2\sqrt{\text{Fo}}} \right)}{\sqrt{\overline{R}_2^2 + (Z + \overline{H})^2}} \right] d\overline{H}$$

$$(7.34)$$

where, as seen in Figure 7.7, $\overline{R}_1$ and $\overline{R}_2$ are dimensionless distances of Boreholes 1 and 2 from coordinate center (0,0).

$$\overline{R}_1 = \frac{R_1}{H} = \frac{\sqrt{(x + h)^2 + y^2}}{H} \quad \text{and} \quad \overline{R}_2 = \frac{R_2}{H} = \frac{\sqrt{(x - h)^2 + y^2}}{H} \tag{7.35}$$

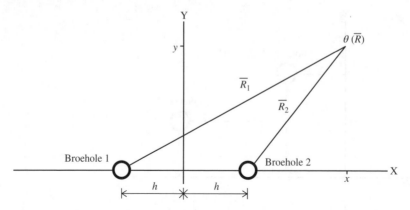

Figure 7.7 System geometric parameters for two boreholes at distances R_1 and R_2 from a fixed point in the surrounding ground.

Note that in using the line-source theory in the case of multiple boreholes, the effects of the boreholes on thermal performance of each other is neglected. Eskilson (1987) determined the error involved in the temperature difference calculated via this method and found it to be negligible.

7.2.2.3 Time Varying Heat Transfer Rates

For time varying heat transfer rates, the problem of heat conduction from the borehole wall to the ground becomes subject to a time-dependent boundary condition $q'(\overline{H}, \tau)$. The variations of heat injection/removal on the borehole can be approximated by a sequence of constant heat fluxes $q'_i(\overline{H})$ where the ith heat flux is applied at $t = \tau_i$ and lasts for a time span Δt_i. Assuming that the governing equations and boundary conditions for the problem are linear, we can obtain the temperature distribution in the body by applying the principle of superposition and obtain the temperature distribution in the body corresponding to the arbitrary continuous boundary condition which we can express as a sequence of, say, n small steps. Therefore, if the temperature rise distribution in the ground corresponding to a constant boundary condition $q'(\overline{H})$ is:

$$\theta(\overline{R}, Z, \text{Fo}) = \int_0^1 q'(\overline{H})I(r, z, t)d\overline{H} \tag{7.36}$$

where

$$I(\overline{R}, Z, \text{Fo}) = \frac{1}{4k\pi} \left[\frac{\text{erfc}\left(\frac{\sqrt{\overline{R}^2 + (Z - \overline{H})^2}}{2\sqrt{\text{Fo}}} \right)}{\sqrt{\overline{R}^2 + (Z - \overline{H})^2}} - \frac{\text{erfc}\left(\frac{\sqrt{\overline{R}^2 + (Z + \overline{H})^2}}{2\sqrt{\text{Fo}}} \right)}{\sqrt{\overline{R}^2 + (Z + \overline{H})^2}} \right]$$

$$\tag{7.37}$$

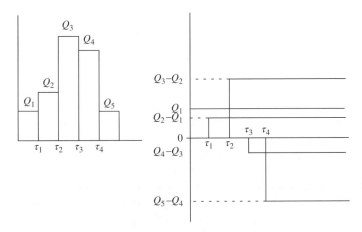

Figure 7.8 Time varying heat transfer rates.

the temperature distribution in the ground corresponding to the varied $q'_i(\overline{H}) = q'(\overline{H}, \tau_i)$ at time t is:

$$
\begin{aligned}
\theta(\overline{R}, Z, \text{Fo}) = &\int_0^1 q'_1(\overline{H}) I(\overline{R}, Z, \text{Fo}_{t-\tau_1}) d\overline{H} \\
&+ \int_0^1 [q'_2(\overline{H}) - q'_1(\overline{H})] I(\overline{R}, Z, \text{Fo}_{t-\tau_2}) d\overline{H} \\
&+ \int_0^1 [q'_3(\overline{H}) - q'_2(\overline{H})] I(\overline{R}, Z, \text{Fo}_{t-\tau_3}) d\overline{H} \\
&+ \dots + \\
&+ \int_0^1 [q'_n(\overline{H}) - q'_{n-1}(\overline{H})] I(\overline{R}, Z, \text{Fo}_{t-\tau_n}) d\overline{H}
\end{aligned}
\tag{7.38}
$$

Here,

$$
\text{Fo}_{t-\tau_i} = \frac{\alpha(t - \tau_i)}{H^2}
\tag{7.39}
$$

and q'_i is the heat flow rate on the line source (here, the borehole) at time τ_i to τ_{i+1}. The concept presented in Equation (7.38) is shown in Figure 7.8. To evaluate the integrations in Equation (7.38) here, a computer code in Fortran is used (Kouhi Fayegh Dekordi 2013).

In the analytical models presented above, a number of assumptions are employed in order to simplify the complicated governing equations. Therefore, the accuracy of the analytical solutions is reduced. The assumptions include treating the two pipes as one pipe coaxial with the borehole or simplifying the pipe and the borehole as an infinitely long line source of heat and not taking into account the thermal capacity inside the borehole. Therefore, regarding time varying heat transfer rates and the influence of surrounding boreholes on both long and short time scales, analytical methods are not as suitable as numerical methods, which are discussed in the next section. However, due to their much shorter computation times, they are still used widely in designing GHEs.

7.3 Numerical Modeling

This section reviews numerical models for HGHEs and VGHEs.

7.3.1 Modeling Vertical Ground Heat Exchangers

System simulation models require the ability to operate at short time scales, often less than a minute. Therefore, the dynamic response of the grout material inside the borehole should be considered. Mei and Baxter (1986) consider the two-dimensional model of the radial and longitudinal heat transfer, which is solved with a finite difference scheme. Yavuzturk (1999) and Yavuzturk *et al.* (1999) present a fully implicit finite volume numerical model based on a two-dimensional radial--axial coordinate system for the simulation of transient heat transfer in vertical ground-loop heat exchangers. Their model is essentially an extension of Eskilson's g-function model to take into account the short-time behavior of the thermal response, for periods of an hour and less. Using an automated parametric grid generation algorithm, numerical grids are generated for various pipe diameters, shank spacings and borehole diameters. Furthermore, the numerical method and grid-generation techniques are validated with a comparable analytical model. Because the short time-step g-function represents the response of the entire GHE, it necessarily utilized a fixed convective resistance. The authors later found it necessary (Yavuzturk and Spitler 2001) to modify the model to include variable convective resistance, but this was done at the expense of modeling the thermal mass of the fluid in the borehole.

Xu and Spitler (2006) describe the development of a new short time-step model for vertical ground-loop heat exchangers. Like the Yavuzturk and Spitler (1999) model, it is an extension to the original long time-step model of Eskilson (1987). However, whereas that model uses a short time-step g-function to account for short time-step effects, the new model replaces the response function approach at short time-steps with a one-dimensional numerical model, which explicitly accounts for the thermal mass of the fluid and the convective resistance as a function of flow rate, fluid mixture, and fluid temperature. This is integrated with Eskilson's long time-step model. By careful control of the one-dimensional model parameters, the model provides acceptably accurate short-term response, without the computational time that is normally required to run such a model continuously throughout the simulation.

Muraya (1995) uses a transient two-dimensional finite element model of the heat transfer around a vertical U-tube heat exchanger for a GSHP system to study the thermal interference between the U-tube legs. He develops two finite element codes: one for pure heat conduction; and another for coupled heat conduction and moisture diffusion. The finite element heat model is coded to approximate solutions to the partial differential heat diffusion equation where no analytical solutions are available. The heat exchanger effectiveness is defined in the model based on ground and grout properties, shank spacings, farfield and loop temperatures, and heat dissipation rates to account for the interference between the U-tube legs. Heat exchanger effectiveness is found to be independent of a dimensionless temperature based on temperatures of the tubes and ground, and varies only with separation distance at steady-state. The model is validated by two analytical cylindrical-source models under constant-temperature and constant-heat flux conditions.

Kavanaugh (1985) uses a two-dimensional finite difference method to describe the performance of a borehole with a concentric tube. Rottmayer *et al.* (1997) uses a two-dimensional finite difference formulation on a polar grid to calculate the lateral heat transfer along 3-m vertical lengths of a borehole. Although axial conduction is neglected, each length section of the model is coupled via the boundary conditions to a model of flow along the U-tube. This quasi-three-dimensional model accounts for the variations in temperature of the borehole fluid in the axial direction.

Lee and Lam (2008) simulate the performance of borehole GHEs using a three-dimensional finite difference method in rectangular coordinates. They evaluate the heat transfer inside the borehole using a finite difference method based on quasi-steady-state conditions, allowing variable temperature and loading along the borehole. Their results show that neither the temperature nor the loading is constant along the borehole, and that the maximum temperature occurs near the top part of the borehole while borehole loading reaches a minimum near the bottom of the borehole. The ground temperature profile changes with distance from the borehole, where the depth of maximum temperature shifts to the mid-level of the borehole at large distances. This implies that using the result obtained from a single borehole with superposition is not sufficient to predict precisely the performance of a borefield. A better approach is to discretize the entire borefield and to simulate all boreholes simultaneously. Finally, the authors compare their results with a line-source solution with superposition and determine that the deviation of their results from the analytical ones increases with the scale of borefield.

Li and Zheng (2009) propose a three-dimensional unstructured finite volume numerical model of a vertical U-tube GHE. They use the Delaunay triangulation method to mesh the cross-section domain of the borefield. The mesh includes both the exterior of the borehole as well as its interior. Thus, the transient effect in the borehole in a short time scale is simulated. To further improve computational accuracy, they divide the ground into several layers in the axial direction, which accounts for the effect of the axial change in temperature. The numerical results of this model agree well with experimental data. This model may be used for simulation of a GHE under any time step size, although its use in transient analysis based on a short time step (an hour or less) is preferred.

Zhongjian and Maoyu (2009) propose a three-dimensional unstructured finite volume numerical model of a vertical U-tube GHE. The Delaunay triangulation method is used to mesh the cross-section domain of the borehole field. The inlet temperature of the borehole fluid is taken as a boundary condition which accounts for the difference (non-symmetry) of heat transfers between the two legs and the surrounding ground, and the conjugate process between them. For further improvement of computational accuracy, the surrounding ground is divided into several layers in the axial direction, to account for the effect of axial temperature change. The numerical results follow experimental data closely.

In a GSHP system, the heat pump and the circulating pumps switch on and off during a given hour; therefore, the effect of thermal mass of the borehole fluid and the dynamics of fluid transport through the loop need to be taken into account. To address this issue, He *et al.* (2009) developed a three-dimensional numerical model, which simulates fluid transport along a pipe loop as well as heat transfer with the ground. The authors carry out the simulation of the GSHP system in EnergyPlus with 10 min time steps for 1 year and use the GLHEPro tool to simulate the borehole heat exchanger. They validate the model by reference to analytical models of borehole thermal resistance and also fluid transport inside the pipe, and compare the predicted outlet temperature with those of a similar two-dimensional model and an implementation of a short time step g-function model. The results show that the delayed response associated with the transit of fluid around the pipe loop is of some significance in moderating swings in temperature during the short period when the heat pump starts to operate. Their BHE model exhibits a lower heat transfer rate over longer periods of operation compared with two-dimensional models. This is due to the mean temperature differences between the fluid and the ground being lower in the three-dimensional model.

Fang *et al.* (2002) consider the variation in load and on–off cycling of the GHE by superposition of a series of heating pulses. The temperature on the borehole wall can then be determined for any instant based on specified operational conditions.

7.3.2 Modeling Horizontal Ground Heat Exchangers

A number of studies have been conducted by various researches in the design, simulation and testing of HGHEs. However, far fewer models are available in the literature for these heat exchangers compared with VGHEs. Several studies of such heat exchangers involve experimentation to determine the effects of varying heat exchanger parameters such as heat flow rate and installation depth, as opposed to mathematical modeling.

Mei (1986) proposes an approach for calculating soil thermal resistance surrounding a HGHE based on the energy balance between the circulating fluid and the surrounding soil. In this approach, soil thermal resistance results from the soil thermal properties and the GHE geometry, and also from the operating strategy of the system. No initial estimation of the soil resistance to heat flow is required. Furthermore, the heat transfer interaction between the borehole fluid and the soil is not assumed, but calculated, based on the inlet water temperature to the GHE and the mass flow rate. Another important difference is that the soil temperature distribution can be directly calculated, which allows for a more accurate prediction of the water temperature profile in the pipe.

A mathematical model of a horizontal type GHE is developed by Piechowski (1999). This model uses elements of the model proposed by Mei (1986). However, some major modifications are made in order to include heat and mass transfer in the soil as well as to enhance the accuracy of the model, and at the same time, to increase the speed of calculations. This is done by concentrating computational effort in the vicinity of the pipe where the most important heat and mass transfer phenomena are taking place. The proposed model calculates the temperature and moisture gradients at the pipe–soil interface.

Tzaferis *et al.* (1992) studied eight models for ground-source air heat exchangers and their algorithms can be classified into two groups:

- Assessment of convective heat transfer from circulating air to the pipe and then conductive heat transfer from the pipe to the ground and in the ground. The necessary input data are geometric characteristics of the system, thermal characteristics of the ground, thermal characteristics of a GHE pipe, and the undisturbed ground temperature during the system operation.
- Assessment of only convective heat transfer from the circulating air to the pipe. The necessary input data are geometric characteristics of the system, thermal characteristics of the pipe, and the temperature of the pipe surface.

De Paepe and Janssens (2003) use a one-dimensional model to analyze the influence of design parameters of the ground-source air heat exchanger on the system's thermohydraulic performance. A relation is derived for the specific pressure drop, linking thermal effectiveness with pressure drop of the air flow inside the tube. Similar algorithms can be applied to predict the performance of a GHE with liquid flow inside a pipe.

Esen *et al.* (2007a) develop a numerical model of heat transfer in the ground for determining the temperature distribution in the vicinity of the HGHE. In their experimental study, they present the coefficient of performance of the GSHP system and the temperature distributions measured in the ground for the 2002–2003 heating season. An analytical solution of the transient temperature response is derived in a semi-infinite medium with a finite length line source of heat theory.

Less time consuming GSHP system simulations and methods of analysis exist, that do not incur excessive sacrifices in accuracy. Such tools provide a balance between these two approaches. An example is a computer simulation package called Ground Heat Exchanger Analysis, Design and Simulation (GHEADS), developed by Tarnawski and Leong (1990). GHEADS is also used for sizing and predicting GSHP performance. With this package, the impacts of low thermal conductivity of volcanic soils on the length of the GHE and long-term use of combined heating and cooling operation on the ground environment (degradation of ground thermal and moisture storage capacity) can be examined (Tarnawski *et al.* 2009). The simulations are based on daily average climatic data in Sapporo, Japan. This system is demonstrated to have a relatively low impact on thermal degradation of the ground environment. As a result, application of horizontal loops for new and retrofit residential and commercial use in northern Japan appears to be feasible.

Leong *et al.* (2006) study the use of a HGHE and the impact of heat deposition and extraction in the ground. They design an optimum GSHP system for an existing dwelling in Ontario using GHEADS. There are many factors to be considered when optimizing the design of a GHE. The optimum design of the GHE in their study may only be applicable to the studied dwelling, because it is specific to the site characteristics (such as soil type and climatic conditions) and system operating parameters (such as magnitude and frequency of heating and/or cooling operation). The optimum GSHP system designed for the existing dwelling appears to have both economic and environmental benefits. GHEADS is used in a modeling example on HGHEs in Section 8.4.

Several experimental studies on HGHEs address various aspects of GSHP systems and verification methods:

- Inalli and Esen (2004) validate the effects of buried depth of a HGHE, and the mass flow rates of a water–antifreeze solution and sewer water on the performance of a GSHP system used for space heating.
- Pulat *et al.* (2009) perform an experimental study on a GSHP and evaluate the performance of a GSHP with HGHEs by considering various system parameters for the winter climatic condition of Bursa, Turkey. For this purpose, a previously used experimental facility for a cooling cycle is modified to a heating cycle. The preliminary numerical temperature distribution around GHE pipes is obtained using the ANSYS finite element software (ANSYS .). The soil thermal conductivity is estimated and compared with the cooling case for reliability purposes; a higher value is used in

numerical heat transfer analysis by considering soil moisture increase due to rainfall. In the analysis, the VGHE cross section, where inlet and outlet water temperatures are measured, is taken into consideration. Numerical solutions are obtained using the finite element method, assuming two-dimensional steady-state conduction.

- Esen *et al.* (2007a) developed a numerical model of heat transfer in the ground for determining the temperature distribution in the vicinity of the HGHE. Neglecting the effect of moisture, it is assumed that heat is transferred mainly by conduction, while the contributions of convection and radiation are negligible. Square finite difference meshes, in the underground field, make the simulation straightforward and less time demanding. The explicit finite difference approximation is used in the numerical analysis. To avoid divergent oscillations in nodal temperatures, the value of time steps must be maintained below a certain upper limit established by the stability criterion. The input data are outdoor air temperature, water–antifreeze solution temperature and mass flow rate of the water–antifreeze solution, which are obtained from experimental measurements. They also carry out an experimental analysis of the system and observe that the numerical results agree with the experimental results.
- In a later study, Esen *et al.* (2007b) determine energy and exergy efficiencies of a GSHP system as a function of trench depth for a heating season. The results show that efficiencies of both systems increase with temperature of the heat source (ground) for a heating sesason. In addition, increasing reference environment temperature reduces the exergy efficiency in both HGHEs.

7.4   Closing Remarks

The modeling of GHEs is described in this chapter. It is shown that the solutions inside the borehole are mostly steady-state, whereas transient effects must be taken into account outside the borehole. Most analytical models for the region outside the borehole do not take into account the thermal capacity of the borehole and assume infinite borehole length (but they are still used widely in designing GHEs because of their much shorter computation times). Analytical models mostly focus on a constant ground heat load and further studies are needed to improve these models for use with transient periodic ground heat loads. The analytical models are able to provide a solution for the area inside the borehole or outside the borehole depending on what is required. In cases where a full system is being studied, including analysis of the inside and outside of the BHE, the literature lacks studies where these analytical models are coupled. When used for sizing GHEs, simplifying assumptions in the line-source theory, such as uniform and constant ground properties, tend to oversize the GHE (i.e., increase its length), thus elevating the system cost and making it less competitive with other conventional heating/cooling systems. Moisture migration does not have a large impact on temperature and heat flow in the ground surrounding vertical heat exchangers. Groundwater flow, when present, is found to have an impact on the heat flow in the ground surrounding a VGHE. However, this impact is often negligible for low rates of groundwater flow. A review of the numerical models shows that the numerical methods are less often applied in modeling larger systems due to their computational time and the required large solution domain.

References

American Water Works Association (AWWA) (2006a) *Contaminant Transport Through Aquitards: A State of the Science Review*, American Water Works Association, AWWA Research Foundation.

American Water Works Association (AWWA) (2006b) *Contaminant Transport Through Aquitards: Technical Guidance for Aquitard Assessment*, American Water Works Association, AWWA Research Foundation.

ANSYS (n.d.) ANSYS FLUENT 12.0 Theory Guide. http://www.afs.enea.it/project/neptunius/docs/fluent/html/th/node359.htm (accessed January 29, 2016).

Bandyopadhyay, G., Gosnold, W. and Mannc, M. (2008) Analytical and semi-analytical solutions for short-time transient response of ground heat exchangers. *Energy and Buildings*, **40** (**10**), 1816–1824.

Bernier, M., Pinel, A., Labib, P. and Paillot, R. (2004) A multiple load aggregation algorithm for annual hourly simulations of GCHP systems. *HVAC&R Research*, **10** (**4**), 471–487.

Bose, J.E., Parker, J.D. and McQuiston, F.C. (1985) *Design/Data Manual for Closed-Loop Ground Coupled Heat Pump Systems*, Oklahoma State University for ASHRAE, Stillwater, OK.

Carslaw, H.S. and Jaeger, J.C. (1946) *Conduction of Heat in Solids*, Claremore Press, Oxford.

Chiasson, A.D., Rees, S.J. and Spitler, J.D. (2000) A preliminary assessment of the effects of groundwater flow on closed-loop ground-source heat pump systems. *ASHRAE Transactions*, **106** (**1**), 380–393.

Claesson, J. and Dunand, A. (1983) *Heat Extraction From the Ground by Horizontal Pipes: A Mathematical Analysis*, Swedish Council for Building Research, Stockholm.

Claesson, J. and Hellström, G. (2011) Multipole method to calculate borehole thermal resistances in a borehole heat exchanger. *HVAC&R Research*, **17**, 895–911.

Cui, P., Yang, H. and Fang, Z.H. (2006) Heat transfer analysis of ground heat exchangers with inclined boreholes. *Applied Thermal Engineering*, **26**, 1169–1175.

De Paepe, M. and Janssens, A. (2003) Thermo-hydraulic design of earth-air heat exchangers. *Energy Buildings*, **35**, 389–397.

Diao, N.R., Li, Q. and Fang, Z.H. (2004a) Heat transfer in ground heat exchangers with groundwater advection. *International Journal of Thermal Sciences*, **43**, 1203–1211.

Diao, N.R., Zeng, H.Y. and Fang, Z.H. (2004b) Improvement in modeling of heat transfer in vertical ground heat exchangers. *HVAC&R Research*, **10** (**4**), 459–470.

Esen, H., Inalli, M. and Esen, M. (2007a) Numerical and experimental analysis of a horizontal ground-coupled heat pump system. *Building and Environment*, **42** (**3**), 1126–1134.

Esen, H., Inalli, M., Esen, M. and Pihtili, K. (2007b) Energy and exergy analysis of a ground-coupled heat pump system with two horizontal ground heat exchangers. *Building and Environment*, **42**, 3606–3615.

Eskilson, P 1987 *Thermal analysis of heat extraction boreholes*. PhD thesis. University of Lund.

Fang, Z.H., Diao, N.R. and Cui, P. (2002) Discontinuous operation of geothermal heat exchangers. *Tsinghua Science and Technology*, **7** (**2**), 194–197.

Gehlin, S.E.A. and Hellström, G. (2003) Influence on thermal response test by groundwater flow in vertical fractures in hard rock. *Renewable Energy*, **28** (**14**), 2221–2238.

Goodrich, L.E. (1980) Three-time-level methods for the numerical solution of freezing problems. *Cold Regions Science and Technology*, **3**, 237–242.

Gu, Y. and O'Neal, D.L. (1998) Development of an equivalent diameter expression for vertical U-tube used in ground-coupled heat pumps. *ASHRAE Transactions*, **104**, 347–355.

He, M, Rees, S & Shao, L 2009 Simulation of a domestic ground source heat pump system using a transient numerical borehole heat exchanger model. *Proceedings of the 11th International Building Performance Simulation Association Conference*, July 27–30, 2009, Glasgow, UK. International Building Performance Simulation Association, pp. 607–614.

Hecht-Mendez, J., Molina-Giraldo, N., Blum, P. and Bayer, P. (2010) Evaluating MT3DMS for heat transport simulation of closed geothermal systems. *Ground Water*, **48** (**5**), 741–756.

Hellström, G 1991 *Ground heat storage: Thermal analyses of duct storage systems*. PhD thesis. University of Lund.

Hromadka, T.V. II,, Guymon, G.L. and Berg, R.L. (1981) Some approaches to modelling phase change in freezing soils. *Cold Regions Science and Technology*, **4**, 137–145.

Inallı, M. and Esen, H. (2004) Experimental thermal performance evaluation of a horizontal ground source heat pump system. *Applied Thermal Engineering*, **24** (**14–15**), 2219–2232.

Jacovides, C.P., Mihalakakou, G., Santamouris, M. and Lewis, J.O. (1996) On the ground temperature profile for passive cooling applications in buildings. *Solar Energy*, **57**, 167–175.

Jun, L., Xu, Z., Jun, G. and Jie, Y. (2009) Evaluation of heat exchange rate of GHE in geothermal heat pump systems. *Renewable Energy*, **34** (**12**), 2898–2904.

Kavanaugh, SP 1985 *Simulation and experimental verification of vertical ground coupled heat pump systems*. PhD thesis. Oklahoma State University.

Kavanaugh, S.P. (1995) A design method for commercial ground-coupled heat pumps. *ASHRAE Transactions*, **101** (**2**), 1088–1094.

Kouhi Fayegh Dehkordi, S 2013 *Thermal sustainability of low-temperature geothermal energy systems: system interactions and environmental impacts*. PhD thesi., University of Ontario Institute of Technology.

Lee, C.K. and Lam, H.N. (2008) Computer simulation of borehole ground heat exchangers for geothermal heat pump systems. *Renewable Energy*, **33** (**6**), 1286–1296.

Leong, W.H., Lawrence, C.J., Tarnawski, V.R. and Rosen, M.A. (2006) Evaluation of a ground thermal energy storage system for heating and cooling of an existing dwelling, in *IBPSA-Canada's 4th Biennial Building Performance Simulation Conference*, University of Toronto, Toronto.

Leong, W.H. and Tarnawski, V.R. (2010) Effects of simultaneous heat and moisture transfer in soils on the performance of a ground source heat pump system, in *ASME-ATI-UIT Conference on Thermal and Environmental Issues in Energy Systems*, Sorrento, Italy.

Leong, W.H., Tarnawski, V.R., Koohi-Fayegh, S. and Rosen, M.A. (2011) Ground thermal energy storage for building heating and cooling, in *Energy Storage* (ed. M.A. Rosen), Nova Science Publishers, New York, pp. 421–440.

Li, Z. and Zheng, M. (2009) Development of a numerical model for the simulation of vertical U-tube ground heat exchangers. *Applied Thermal Engineering*, **29** (**5–6**), 920–924.

Mei, VC 1986 *Horizontal Ground-Coupled Heat Exchanger, Theoretical and Experimental Analysis*. Oak Ridge National Laboratory: Technical Report, ORNL/CON-193.

Mei, V.C. and Baxter, V.D. (1986) Performance of a ground-coupled heat pump with multiple dissimilar U-tube coils in series. *ASHRAE Transactions*, **92 (2)**, 22–25.

Mihalakakou, G. (2002) On estimating ground surface temperature profiles. *Energy and Buildings*, **34**, 251–259.

Mihalakakou, G. and Lewis, J.O. (1996) The influence of different ground covers on the heating potential of the earth-to-air heat exchangers. *Renewable Energy*, **7**, 33–46.

Muraya, NK 1995 *Numerical modeling of the transient thermal interference of vertical U-tube heat exchangers*. PhD thesis. Texas A&M University.

Nam, Y., Ooka, R. and Hwang, S. (2008) Development of a numerical model to predict heat exchange rates for a ground-source heat pump system. *Energy and Buildings*, **40 (12)**, 2133–2140.

Paul, ND 1996 *The effect of grout conductivity on vertical heat exchanger design and performance*.,Master Thesis. South Dakota State University.

Philip, J.R. and de Vries, D.A. (1957) Moisture movement in porous materials under temperature gradients. *Transactions American Geophysical Union*, **38**, 222–232.

Piechowski, M. (1999) Heat and mass transfer model of a ground heat exchanger: Theoretical development. *International Journal of Energy Research*, **23**, 571–588.

Pulat, E., Coskun, S., Unlu, K. and Yamankaradeniz, N. (2009) Experimental study of horizontal ground source heat pump performance for mild climate in Turkey. *Energy*, **34**, 1284–1295.

Rottmayer, S.P., Beckman, W.A. and Mitchell, J.W. (1997) Simulation of a single vertical U-tube ground heat exchanger in an infinite medium. *ASHRAE Transactions*, **103 (2)**, 651–658.

Salah El-Din, M.M. (1999) On the heat flow into the ground. *Renewable Energy*, **18**, 473–490.

Tarnawski, V.R. (1982) An Analysis of Heat and Moisture Movement in Soils in the Vicinity of Ground Heat Collectors for Use in Heat Pump Systems, in *Acta Polytechnica Scandinavica, Mechanical Engineering Series*, vol. **82**, Helsinki: Saint Mary's University, Division of Engineering.

Tarnawski, V.R. and Leong, W.H. (1990) *Computer Simulation of Ground Coupled Heat Pump Systems*, Technical Report, National Research Council, Ottawa.

Tarnawski, V.R., Leong, W.H., Momose, T. and Hamada, Y. (2009) Analysis of ground source heat pumps with horizontal ground heat exchangers for northern Japan. *Renewable Energy*, **34**, 127–134.

Tzaferis, A., Liparakis, D., Santamouris, M. and Argiriou, A. (1992) Analysis of the accuracy and sensitivity of eight models to predict the performance of earth-to-air heat exchangers. *Energy Buildings*, **18**, 35–43.

Xu, X. and Spitler, J.D. (2006) Modeling of vertical ground loop heat exchangers with variable convective resistance and thermal mass of the fluid. *Proceedings of Ecostock 2006*, May 31–June 2, 2006, Stockton, USA. Pomona.

Yang, H., Cui, P. and Fang, Z. (2010) Vertical-borehole ground-coupled heat pumps: A review of models and systems. *Applied Energy*, **87 (1)**, 16–27.

Yang, W., Shi, M., Liu, G. and Chen, Z. (2009) A two-region simulation model of vertical U-tube ground heat exchanger and its experimental verification. *Applied Energy*, **86**, 2005–2012.

Yavuzturk, C. (1999) *Modeling of vertical ground loop heat exchangers for ground source heat pump systems*. PhD thesis. Oklahoma State University.

Yavuzturk, C. and Spitler, J. (1999) A short time step response factor model for vertical ground loop heat exchangers. *ASHRAE Transactions*, **105**, 475–485.

Yavuzturk, C. and Spitler, J.D. (2001) Field validation of a short time step model for vertical ground-loop heat exchangers. *ASHRAE Transactions*, **107** (**1**), 617–625.

Yavuzturk, C., Spitler, J.D. and Rees, S.J. (1999) A transient two-dimensional finite volume model for the simulation of vertical U-tube ground heat exchangers. *ASHRAE Transactions*, **105** (**2**), 465–474.

Zeng, H.Y., Diao, N.R. and Fang, Z. (2002) A finite line-source model for boreholes in geothermal heat exchangers. *Heat Transfer Asian Research*, **31** (**7**), 558–567.

Zeng, H.Y., Diao, N.R. and Fang, Z. (2003a) Heat transfer analysis of boreholes in vertical ground heat exchangers. *International Journal of Heat and Mass Transfer*, **46** (**23**), 4467–4481.

Zeng, H.Y., Diao, N.R. and Fang, Z. (2003b) Efficiency of vertical geothermal heat exchangers in ground source heat pump systems. *Journal of Thermal Science*, **12** (**1**), 77–81.

Zhongjian, L. and Maoyu, Z. (2009) Development of a numerical model for the simulation of vertical U-tube ground heat exchangers. *Applied Thermal Engineering*, **29**, 920–924.

8

Ground Heat Exchanger Modeling Examples

In this chapter, analytical and numerical models presented in the previous chapter are applied to several examples. The examples involve various levels of difficulty and detail. Since the modeling approach for a ground heat exchanger (GHE) problem is often selected based on the modeling objective, the objectives of modeling the systems in each of the examples are stated at the start of the example. The challenges of modeling and the shortcomings of each model example based on its simplifying assumptions are also described. In many of the examples, the modeling techniques introduced in Chapter 7 are used and the definitions of the terms that appear in the relations included in those examples can be found in Chapter 7. The examples considered include the semi-analytical modeling of two boreholes, the numerical modeling of two boreholes, the numerical modeling of a borefield, and the numerical modeling a horizontal ground heat exchanger (HGHE).

8.1 Semi-Analytical Modeling of Two Boreholes

Although a number of studies have focused on the development and application of ground-source heat pump (GSHP) systems, further investigation is needed, particularly in the area of estimating the heat exchanger's heat delivery/removal strength when the ground surrounding them experiences a temperature rise or drop. An example of such case is when thermal interaction among neighbor boreholes occurs and the efficiency of the heat pump is affected. It is important to develop and utilize models that account for the drop in heat delivery strength when the borehole wall temperature increases during the operation time or by another nearby operating system. Such a model needs to be capable of monitoring how the increase in the surrounding ground affects the working fluid temperature needed to deliver/extract heat to/from the ground at a given rate. An example modeling approach for such analysis is presented in this section.

8.1.1 Physical Domain

A domain consisting of two vertical borehole heat exchangers having a distance of D_b from each other is considered (Figure 8.1a and b). The borehole fluid runs through a U-tube (Figure 8.1c) and delivers or removes heat through the grout in the borehole and the ground surrounding it.

Geothermal Energy: Sustainable Heating and Cooling Using the Ground, First Edition.
Marc A. Rosen and Seama Koohi-Fayegh.
© 2017 John Wiley & Sons, Ltd. Published 2017 by John Wiley & Sons, Ltd.

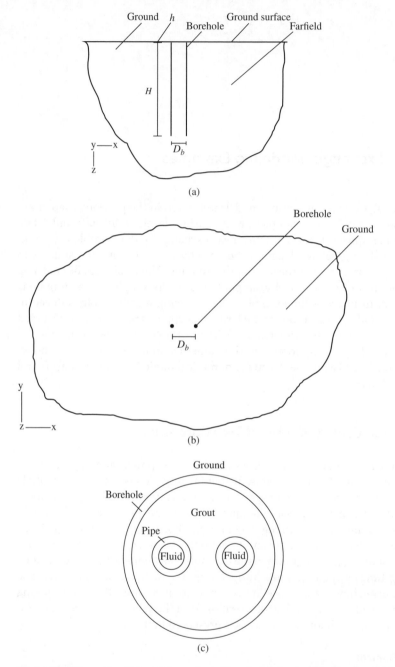

Figure 8.1 Schematic of (a) *xz* cross section of two boreholes installed at a borehole separation distance D_b, (b) *xy* cross section of two boreholes installed at a certain borehole distance, and (c) cross section of inside a borehole.

8.1.2 Assumptions

Modeling a borehole heat exchanger completely by accounting for all the varying parameters affecting the borehole heat exchange operation is challenging. The current model is developed based on a set of simplifying assumptions. Some of the simplifying assumptions are made due to a lack of experimentally evaluated physical parameters or due to the ability of the modeling tools available while others are made because of the negligible effects of a parameter in the model. The following assumptions are made here, and are divided into two groups.

First, the following assumptions are made due to a lack of experimentally evaluated physical parameters or the limitations in available modeling tools that affect simulation results:

- Thermal properties of the ground, grout, and the borehole fluid are isotropic and uniform.
- The impact of groundwater advection is negligible.

The thermal properties of the ground such as thermal conductivity, specific heat, and density vary with ground composition. Along the length of a typical vertical ground heat exchanger (VGHE), the ground surrounding the boreholes can have different layers with variable thicknesses and, therefore, properties. However, in this example, it is assumed that the thermal properties of the ground are constant in order to simplify the modeling.

When groundwater flow is present, the heat flow rate in the ground surrounding the borehole tends to increase in the direction of the flow. However, for this example it is assumed that groundwater flow is not present to simplify the modeling.

Secondly, the following assumptions are invoked, but are expected to have only minor impacts on the validity of simulation results of the current example:

- Moisture migration is negligible.
- The dominant mode of heat transfer in the ground and the grout is conduction.
- The dominant mode of heat transfer in the borehole fluid region is convection.
- There is no thermal energy generation in any of the regions.
- The borehole fluid is incompressible and the pressure variation is neglected.
- The ground surface is assumed to be isothermal and time invariant during system operation.
- The thermal resistances of the borehole wall and the pipe are neglected.
- The contact resistances between the borehole wall and the ground and between the borehole wall and the grout are neglected.

8.1.3 Method

In order to determine heat delivery/removal strength of the borehole fluid, the borehole wall temperature must be defined by coupling the heat transfer model inside the borehole to the one outside the borehole via the borehole wall temperature or the heat flow rate per unit length of borehole.

Two analytical approaches are used to calculate the temperature profiles in the ground surrounding boreholes as well as inside the borehole. To model the heat transfer inside the borehole, the model of Zeng *et al.* (2003) (see Section 7.2.2) can be used to formulate the temperature profiles of the fluids flowing in the U-tubes in the boreholes. The heat transfer outside the borehole is modeled by modifying the semi-analytical model of Zeng et al. (2002, 2003) (see Section 7.2.2.3), which evaluates the temperature in the ground surrounding a borehole, by using a temporal superposition method that is able to estimate these temperatures when the ground heat load is transient. A coupling procedure is presented in this section that uses both models inside the borehole and outside the borehole in order to relate the temperature variations in the ground surrounding the borehole to the corresponding temperature variations in the working fluid inside the borehole. How the two models are coupled and what parameters are kept constant vary depending on the objective of the study.

8.1.3.1 Model Coupling via Heat Flow Rate

The heat transferred to the ground from each of the pipes in the borehole can be obtained as:

$$q'(z) = \frac{T_{f1}(z) - T_b}{R_1^{\Delta}} + \frac{T_{f2}(z) - T_b}{R_2^{\Delta}} \tag{8.1}$$

where $R_1^{\Delta} = R_2^{\Delta}$ for a symmetric U-tube configuration. Using the dimensionless parameters introduced in Equation (7.22), Equation (8.1) can be rewritten in terms of several dimensionless parameters:

$$q'(z) = (T_f' - T_b) \left[\frac{\Theta_1(Z)}{R_1^{\Delta}} + \frac{\Theta_2(Z)}{R_2^{\Delta}} \right] \tag{8.2}$$

This is the spatial distribution of the heating strength along the rod (Figure 8.2). In order to compare the results obtained with a constant heat flux model with the results from this variable heat source model, an equivalent inlet temperature (T_f') for the variable heat source model, resulting in the same total heat conduction in the ground, can be assumed. The total heat flow rate is calculated by integrating the heat flow rate along the borehole:

$$\frac{Q}{H} = \int_0^1 q'(Z)dZ = \int_0^1 (T_f' - T_b) \left[\frac{\Theta_1(Z)}{R_1^{\Delta}} + \frac{\Theta_2(Z)}{R_2^{\Delta}} \right] dZ \tag{8.3}$$

Assuming a constant borehole wall temperature and inlet fluid temperature, Equation (8.3) can be used to calculate the equivalent inlet fluid temperature in the variable heat source model that results in the same heat delivery/removal to the surrounding ground. That is,

$$T_f' = T_b + \frac{\dfrac{Q}{H}}{\displaystyle\int_0^1 \left[\frac{\Theta_1(Z)}{R_1^{\Delta}} + \frac{\Theta_2(Z)}{R_2^{\Delta}} \right] dZ} \tag{8.4}$$

Note that the integration in the denominator of Equation (8.4) depends on only the geometric specifications of the borehole and thermal characteristics of the grout.

In ground heat delivery, the heating strength of the variable heat source declines along the borehole as the borehole fluid temperature decreases, by losing heat to the grout and then the ground. In ground heat removal, the heat removal strength of the variable heat source declines along the borehole as the borehole fluid temperature increases by

Figure 8.2 Distribution of heat flux along the borehole length.

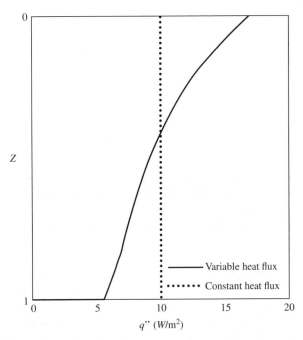

gaining heat from the grout and then the ground. This heat flow rate [Equation (8.2)] can be used in the model outside the borehole [Equation (7.30)] to model the temperature variations in the ground due to a variable heat source. The variation of the variable heating strength along the borehole length is shown in Figure 8.2.

The relations derived in this section are able to show the existence of a variable heating strength due to the variable temperature of the borehole fluid along the borehole wall. In the variable heat source model, certain simplifying assumptions such as constant ground temperature are made. When calculating the heat input to the ground, it becomes clear that it varies with the borehole wall temperature. This assumption ignores the drop in heat injection strength when the borehole wall temperature increases and, therefore, underestimates the inlet temperature of the borehole fluid that is required to meet the heat injection needs of the system. In cases of multiple boreholes, superposition of the temperature excesses resulting from individual boreholes seems to be the most common solution in analytical approaches. In numerical approaches, the boundary condition that plays the role of heat delivery/removal is a heat flow rate per unit length boundary type that, regardless of being constant or variable based on the building needs, does not reflect the drop in the heat injection/removal strength when temperature of the ground surrounding the borehole increases/decreases by its own performance or another nearby system's performance. This assumption forces the system to deliver a desired amount of heat to the ground regardless of the ground temperature. In reality, the amount of heat delivered to the ground is driven by the temperature difference between the borehole fluid and the ground. In some cases, the assumption of constant borehole wall temperature is acceptable considering how the conduction problem is simplified. However, when determining how thermal interaction between two operating GHEs can affect their performance, the effect of the transient borehole wall temperature on the heat delivery strength and inlet fluid temperature becomes an important factor. Therefore, the current model is only valid for low temperature variations in the ground surrounding the boreholes which is only achieved by assuming lower heat flux values on the borehole wall. Modifying the current model to one with typical industrial values for GSHP systems requires the ground temperature to be treated as variable. This complicates the coupling procedure and increases the computation time for evaluating the borehole fluid temperature as well as borehole wall temperature. The reader is refered to Appendix B (Section B.3) for illustrations of the temperature response in the ground surrounding a borehole using the model presented in this section.

8.1.3.2 Model Coupling via Borehole Wall Temperature

To evaluate the effect of the temperature increase in the ground surrounding a borehole on the operation of the heat pump for a given heat flow rate, the outlet temperature of the borehole fluid is the key parameter that couples the ground response with the heat pump model. This temperature can be calculated by coupling the two models inside and outside the borehole. In this model, in order to maintain the required heat flow rate as the temperature of the borehole wall increases/decreases over time, the inlet temperature of the borehole fluid is updated to an increased/decreased value in order to deliver/remove the required heat to/from the ground. With the increasing/decreasing inlet fluid temperature, outlet temperature of the borehole fluid is modified accordingly. Monitoring the outlet temperature of the borehole fluid is advantageous since it is the coupling parameter between the model inside the borehole and the heat pump. In addition, this temperature is often the key parameter in the system for determining if the heat pump will operate under the ground temperature conditions.

In order to formulate the process, the solution to the model inside the borehole [Equation (7/21)] at $Z = 0$ is used to calculate the borehole fluid outlet temperature:

$$T_{f,out} = T_{f2}(0) = T_b + (T'_f - T_b)\,\Theta_2(0) \tag{8.5}$$

where $\Theta_2(0)$ varies with system parameters that are known to the system designer or simulator [Equation (7.21)] and

$$\Theta_2(0) = \frac{\cosh(\beta) - \sqrt{\dfrac{1-P}{1+P}}\,\sinh(\beta)}{\cosh(\beta) + \sqrt{\dfrac{1-P}{1+P}}\,\sinh(\beta)} \tag{8.6}$$

It is seen in Equation (8.5) that the outlet temperature of the borehole fluid varies with the borehole wall temperature T_b and the borehole inlet fluid temperature T'_f. Another correlation between the outlet fluid temperature and the borehole inlet fluid temperature can be written as:

$$T_{f,out} = T'_f - \frac{q'_{ave}H}{\dot{m}c_p} \tag{8.7}$$

where q'_{ave} is the average heat flow rate per unit length of the borehole. Note that $q'_{ave}H$ in Equation (8.7) is the total amount of heat delivered to/removed from the ground. Using Equation (8.5) and Equation (8.7), one can determine the inlet and outlet temperatures of the borehole fluid that are needed to deliver/remove a required amount of heat to/from the ground at a given borehole wall temperature. Thus,

$$T_f' = T_b - \frac{q_{ave}'H}{\dot{m}c_p[1 - \Theta_2(0)]}$$

$$T_{f,out} = T_b - \frac{q_{ave}'H\,\Theta_2(0)}{\dot{m}c_p[1 - \Theta_2(0)]} \tag{8.8}$$

As mentioned previously, the borehole wall temperature and the borehole heat flow rate are the coupling parameters between the model inside the borehole and the model outside the borehole. The borehole heat flow rate can be substituted in the model outside the borehole [Equation (7.30)], or its modified versions [Equation (7.32) and Equation (7.38)], and the borehole wall temperature can be evaluated accordingly. In order to use the solution to the line-source theory in evaluation of the borehole wall temperature, Equation (7.30) is used for $\overline{R} = r_b/H$. This may cause a small error in the calculated value for the borehole wall temperature since using the line-source model outside the borehole at $\overline{R} = r_b/H$ assumes thermal properties of the ground for regions smaller than $(\overline{R} < r_b/H)$ whereas, in reality, the grout and borehole fluid with different thermal properties are present in $\overline{R} < r_b/H$.

The temperature of the borehole wall varies along the borehole length as shown in Equation (7.30). However, in the derivation of the solution to the model for inside the borehole [Equation (7.21)], a constant borehole wall temperature along the borehole length is assumed. Thus, Equation (8.8) is only valid for cases where the borehole wall temperature is assumed constant or its average is used. In Equation (7.30), the variation of the borehole wall temperature along the borehole length is so small that the value of the borehole wall temperature at $Z = 0.5$ can be used as a good estimate of the average borehole wall temperature. Integrating this value along the borehole length is another alternative that is not preferred due to its computation time. A comparison of the two options (Figure 8.3) confirms the accuracy of using the borehole wall temperature in mid-length of the borehole as a good estimate of average borehole wall temperature.

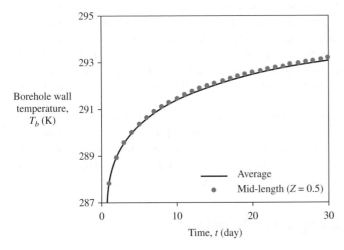

Figure 8.3 Comparison of average borehole wall temperature along borehole length and borehole wall temperature at borehole mid-length ($Z = 0.5$).

Thus, the borehole wall temperature is calculated for any time after the start of system operation as follows:

$$T_b(\text{Fo}) = T_0 + \frac{1}{4k\pi} \int_0^H q'(\overline{H})$$

$$\times \left\{ \frac{\text{erfc}\left(\dfrac{\sqrt{\overline{R}_b^2 + (0.5 - \overline{H})^2}}{2\sqrt{\text{Fo}}} \right)}{\sqrt{\overline{R}_b^2 + (0.5 - \overline{H})^2}} - \frac{\text{erfc}\left(\dfrac{\sqrt{\overline{R}_b^2 + (0.5 + \overline{H})^2}}{2\sqrt{\text{Fo}}} \right)}{\sqrt{\overline{R}_b^2 + (0.5 - \overline{H})^2}} \right\} d\overline{H} \qquad (8.9)$$

For time-varying heat transfer rates, the borehole wall temperature can be calculated from Equation (7.38) by substituting the value of borehole radius R_b and $Z = 0.5$. Hence,

$$T_b(\text{Fo}) = T_0 + \int_0^1 q_1'(\overline{H})\, I(\overline{R}_b, 0.5, \text{Fo}_{t-\tau_1})\, d\overline{H}$$

$$+ \int_0^1 [\, q_2'(\overline{H}) - q_1'(\overline{H})]\, I(\overline{R}_b, 0.5, \text{Fo}_{t-\tau_2})\, d\overline{H}$$

$$+ \int_0^1 [\, q_3'(\overline{H}) - q_2'(\overline{H})]\, I(\overline{R}_b, 0.5, \text{Fo}_{t-\tau_3})\, d\overline{H} \qquad (8.10)$$

$$+ \ldots +$$

$$+ \int_0^1 [\, q_n'(\overline{H}) - q_{n-1}'(\overline{H})]\, I(\overline{R}_b, 0.5, \text{Fo}_{t-\tau_n})\, d\overline{H}$$

The accuracy of the solution can be improved by increasing the number of the time steps. Here, the variations in the ground heat load profile are assumed to be known according to the building needs and the heat pump operation. The borehole wall temperature can be updated at every time step [Equation (8.10)] in order to estimate the inlet and borehole outlet fluid temperatures [Equation (8.8)]. However, evaluating the borehole wall temperature in the presence of time varying heat flow rates [Equation (8.10)] becomes computationally intensive and requires efficient algorithms that lower the number of time steps (Yavuzturk and Spitler 1999; Bernier *et al.* 2004; Marcotte and Pasquier 2008).

In this example, the model available for outside the borehole is used as a basis for calculating the borehole wall temperature. The procedure for coupling this temperature with the model available for inside the borehole to calculate the inlet and borehole outlet fluid temperature according to the variable borehole wall temperature is also presented in this chapter. Figure 8.4 summarizes the modeling procedure.

8.2 Numerical Modeling of Two Boreholes

In this section, two boreholes are modeled using a three-dimensional domain representing the ground surrounding the boreholes as well as the grout and the borehole

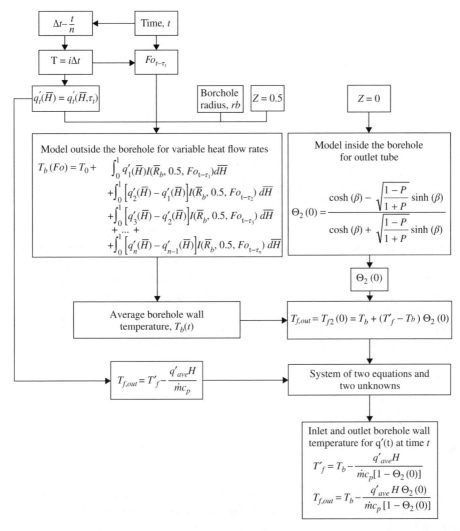

Figure 8.4 Coupling procedure for borehole wall temperature and the model for inside the borehole to calculate the inlet and borehole outlet fluid temperatures according to the variable borehole wall temperature.

fluid inside the boreholes. The objective of such modeling could be to determine how a temperature rise caused by one borehole (or borehole system) can affect the operation of a neighboring borehole (or borehole system). This section ends with a description of the challenges related to such modeling. Note that the problem and its objectives are similar to those in the previous section, but a different modeling approach is employed.

8.2.1 Physical Domain

The physical domain presented in Section 8.1.1 is modeled here (Figure 8.1a and b) using a numerical solution. The borehole fluid runs through a U-tube (Figure 8.1c) and delivers or removes heat to its surrounding, which is grout in the borehole and the ground surrounding the borehole.

8.2.2 Governing Equations

In the current problem, general forms of the continuity, momentum and energy equations for an incompressible fluid can be written as:

$$\nabla \cdot \vec{v} = 0 \tag{8.11}$$

$$\rho \left(\frac{\partial \vec{v}}{\partial t} + \nabla \vec{v} \cdot \vec{v} \right) = -\nabla p + \mu \nabla^2 \vec{v} + F \tag{8.12}$$

$$\rho c_p \left(\frac{\partial T}{\partial t} + \nabla T \cdot \vec{v} \right) = k \nabla^2 T + \mu \phi \tag{8.13}$$

where $\vec{v}$, T, p, ρ, μ, k, and $\mu\phi$ are the flow velocity vector, temperature, static pressure, density, molecular viscosity, conductivity, and viscous dissipation, respectively. The two terms on the right-hand side of Equation (8.13) represent energy transfer due to conduction and viscous dissipation, respectively, while the terms on the left-hand side of this equation represent energy change and energy transfer due to convection.

Heat transfer in the ground and the grout is in the form of conduction while the dominant mode of heat transfer in the borehole fluid region is convection. Therefore, the ground, grout and borehole fluid regions are presented separately.

8.2.3 Borehole Fluid Region

The dominant mode of heat transfer in the borehole fluid region is convection. All the above governing equations are applied in this region to evaluate the borehole fluid temperature T_f.

An initial temperature (equal to the undisturbed ground temperature), T_0, is assumed:

$$T_f = T_0 \quad \text{at} \quad t = 0 \tag{8.14}$$

The borehole fluid model uses the inlet fluid temperature as its boundary condition to evaluate the temperature of the borehole fluid along the borehole as well as its outlet temperature:

$$T_f|_{z=0} = T_f'(t) \tag{8.15}$$

where T_f' is the borehole fluid inlet temperature, which varies with time. At the inlet of the U-tube, the momentum equation has the following boundary condition:

$$\vec{v}|_{z=0} = -v_0 \hat{k} \tag{8.16}$$

where v_0 is the inlet velocity of the borehole fluid.

It is assumed that for a certain length (h) at the top of the U-tube that the tube wall has an adiabatic condition. Therefore,

$$\frac{\partial T_f}{\partial r^*} = 0 \quad 0 \leq z \leq h \tag{8.17}$$

where r^* is the direction perpendicular to the U-tube surface.

8.2.4 Grout Region

Based on the assumptions presented in Section 8.1.2, heat transfer in the grout region is in the form of conduction and the energy equation for this region [Equation (8.13)] reduces to:

$$\frac{1}{\alpha}\frac{\partial T_g}{\partial t} = \nabla^2 T_g \tag{8.18}$$

An initial temperature (equal to the undisturbed ground temperature), T_0, is assumed:

$$T_g = T_0 \text{ at } t = 0 \tag{8.19}$$

The model for the grout area simply relates the temperature of the borehole fluid T_f to the borehole wall temperature T_b. The temperature of the grout at the U-tube can be used as a coupled boundary condition to the temperature of the borehole fluid. Therefore,

$$T_g|_{\text{at the tube wall}} = T_f|_{\text{at the tube wall}} \qquad z \geq h \tag{8.20}$$

where T_f can be calculated via the model for the borehole fluid described in Section 8.2.3. The grout model takes a uniform temperature (equal to the undisturbed ground temperature) at the ground surface. Therefore,

$$T_g|_{z=0} = T_0 \tag{8.21}$$

where T_0 is the undisturbed ground temperature.

8.2.5 Ground Region

Based on the assumptions presented in Section 8.1.2, heat transfer in the ground region is in the form of conduction and the energy equation [Equation (8.13)] reduces to:

$$\frac{1}{\alpha}\frac{\partial T_s}{\partial t} = \nabla^2 T_s \tag{8.22}$$

An initial temperature (equal to the undisturbed ground temperature), T_0, is assumed:

$$T_s = T_0 \quad \text{at} \quad t = 0 \tag{8.23}$$

The ground model takes a uniform temperature (equal to the undisturbed ground temperature) at the ground surface and at the farfield:

$$T_s|_{z=0} = T_0 \tag{8.24}$$

$$T_s|_{\text{at farfield}} = T_0 \tag{8.25}$$

Note that the farfield temperature boundary condition is only used for numerical simulation of the system. This temperature is assumed to be equal to the ground initial temperature and completes the model. However, this temperature is not supposed to affect the temperature of the ground surrounding the borehole that is calculated during the simulation. Therefore, this boundary can be switched to an adiabatic boundary.

The ground at the bottom of the borehole takes the same temperature as the grout at the bottom of the borehole:

$$T_s|_{z=H+h} = T_g|_{z=H+h} \qquad 0 \leq r^{**} \leq r_b \tag{8.26}$$

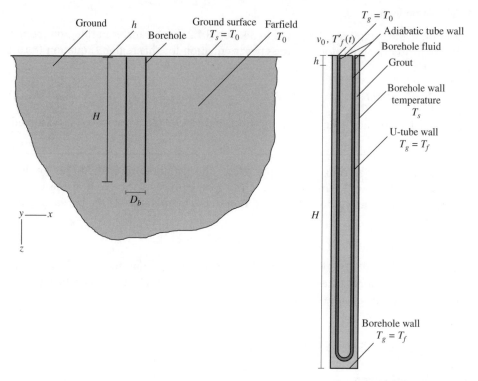

Figure 8.5 Boundary conditions in the example for two boreholes separated by a distance D_b.

where r^{**} is the radial distance from the borehole axis and r_b is the borehole radius. In a similar manner, at the borehole wall, when coupling this boundary to the model inside the borehole, the temperature of the borehole wall calculated from the grout model can be used to evaluate the temperature of the ground surrounding the borehole. Therefore,

$$T_s|_{r^{**}=r_b} = T_g|_{r^{**}=r_b} \qquad h \le z \le H + h \tag{8.27}$$

Note that when only the heat flow in the ground is of concern, the grout and the borehole fluid regions are not included in the problem setup and the boundary condition in Equation (8.27) is modified to a heat boundary condition from the borehole wall to the ground:

$$-k\frac{\partial T_s}{\partial r}\bigg|_{r^{**}=r_b} = q''(t) \tag{8.28}$$

The boundary conditions that are applied to the numerical model are summarized in Figure 8.5.

8.2.6 Physical Parameters and Geometric Specifications

Physical parameters and geometric specifications that are used in the model are described in this section. The model input parameters and properties are taken from various sources and included in Table 8.1 (Shonder and Beck 1999; Incropera and DeWitt 2000; Hepbasli *et al.* 2003; Gao *et al.* 2008).

Table 8.1 Physical properties and system specifications.

Parameter	Value
Ground	
Undisturbed ground temperature	9 °C (282 – 283 K)
Ground thermal conductivity	1.5 W/m K
Ground specific heat capacity	1550 J/kg K
Ground density	1950 kg/m^3
Grout	
Grout thermal conductivity	2.6 W/m K
Grout specific heat capacity	1250 J/kg K
Grout density	1600 kg/m^3
Borehole fluid[a]	
Borehole fluid thermal conductivity	0.0242 W/m K
Borehole fluid specific heat capacity	4182 J/kg K
Borehole fluid density	998.2 kg/m^3
Borehole fluid mass flow rate	0.225 kg/s
Borehole geometry	
Total borehole length, H	50 m
Borehole radius, r_b	0.050 m
U-tube radius, r_p	0.016 m
U-tube center-to-center half distance, D	0.026 m

a) As a simplifying assumption, properties of water at atmospheric pressure are used.
Source: Incropera and DeWitt, 2000; Gao *et al.* 2008; Hepbasli *et al.* 2003; Shonder and Beck 1999.

8.2.7 Numerical Solver

Many of the numerical algorithms currently used in simulating heat transfer and fluid mechanics problems are available in ANSYS FLUENT (ANSYS n.d.). In some cases, an algorithm used in one type of problem is not considered a prudent selection in another problem. Incorrect selection of numerical algorithms when setting up a model in ANSYS FLUENT can result in longer computation times and, in some cases, may lead to a solution not converging. This is especially important in modeling larger solution domains or ones containing transient boundary conditions such as those in the current example. Therefore, when using ANSYS FLUENT as a solver for the specific conditions of each problem, the user must know the details of each of the algorithms that are selected and decide which is best suited to the problem.

In this section, the methods selected for simulating the current example are presented. A pressure-based solver in ANSYS FLUENT is used to present the numerical model to solve the governing equations in ground, grout and borehole fluid in a three-dimensional domain (Figure 8.6).

8.2.8 Grid

To start the solution, a control-volume-based technique is used that divides the domain into discrete control volumes using computational grids. Structured curvilinear grids have proven to have difficulty in achieving viable mapping when geometry becomes complex. In these cases, it is often advantageous to subdivide the flow domain into several subregions or blocks, each of which is meshed separately and joined up correctly with its neighbors. For more complex geometries, more blocks are used up to the point where each individual mesh is treated as a block, resulting in the so-called unstructured grid. This gives unlimited geometric flexibility and allows the most efficient use of computing resources. The advantage of such an arrangement is that no implicit structure of coordinate lines is imposed by the grid – hence the name unstructured – and the mesh can be easily concentrated where necessary without wasting computer storage. Moreover, control volumes may have any shape and there are no restrictions on the number of adjacent cells meeting at a point (two-dimensional) or along a line (three-dimensional).

8.2.8.1 Grid Formation

Generating the mesh in the current example is performed in GAMBIT. Another advantage of choosing an unstructured mesh is that it allows the calculation of heat flows in or around the borehole without having to spend a long time on mesh generation and mapping. Grid generation is fairly straightforward with triangular grids and mesh refinement and adaption to improve resolution in regions with large gradients are much easier in unstructured triangular meshes. In the current model, since the multiple borehole geometry does not fit into Cartesian or cylindrical coordinates, an unstructured mesh with triangular and triangular prism elements is chosen for the three-dimensional geometry (Figure 8.7). The vertical section domain may be discretized using structured grids due to relatively uniform vertical structure, as shown in Figure 8.8. The governing equations on the individual control volumes are integrated to construct algebraic equations for the discrete dependent variable, that is, temperature. The discretized equations are linearized and solved to yield updated values of the dependent variables (Versteeg and Malalasekera 2007).

One of the disadvantages of numerical approaches is their computation time for long-term system performance. The diameters of the U-tubes in the borehole are fairly small, on the order of 10^{-2} m, while the size of the solution domain, which depends on the duration of system operation and its heating/cooling load, is approximately on an order of 10 m, making the domain extremely disproportionate. As a result, a large number of mesh elements is required for simulation of a single borehole and its surrounding ground. To achieve an inaccuracy of 2% or less for the steady-state heat transfer analysis of boreholes, a minimum number of approximately 18 elements describing any circular shape of a horizontal cross section is needed (Bauer *et al.* 2011). In modeling the ground surrounding the borehole, a domain of a certain size can work well for one model, while it can be too small for another model requiring more boreholes, or longer system performance durations or higher heating injection/removal rates. At the outer edge of the domain, a constant farfield temperature condition equal

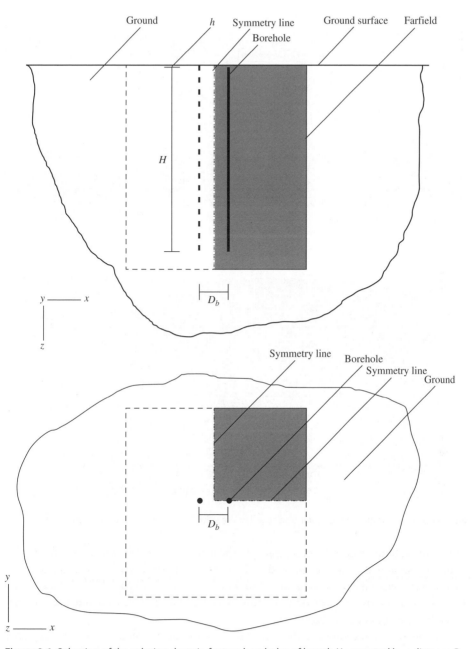

Figure 8.6 Selection of the solution domain for two boreholes of length *H* separated by a distance D_b. The gray area is selected for modeling.

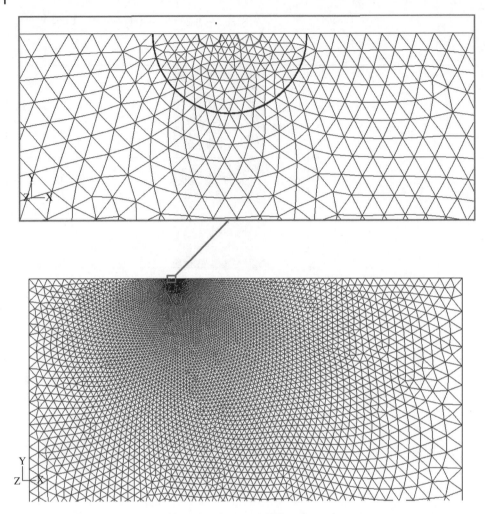

Figure 8.7 Computational triangular grids used in the solution domain in *xy* cross section.

to the initial temperature is often applied. The sensitivity of the solution results to this boundary should always be examined and avoided by increasing the size of the domain. In three-dimensional modeling of a borehole system with typical flow velocities, a vertical element size of 2 m or less often needs to be applied to avoid inaccuracies exceeding 2% (Bauer *et al.* 2011).

To reduce computation time, the heat transfer symmetry about the two vertical planes shown in Figure 8.6 is exploited. Therefore, only a quarter of the borehole field is modeled and the solution domain (ground) is enclosed by the farfield, the ground surface and two symmetry planes. Theoretically, an adiabatic wall boundary condition is replaced on the symmetry line. In Figure 8.6, the gray area is the solution domain, the results of which can be replicated to the other areas drawn with dashed lines due to their symmetry. In addition, the temperature gradient in the domain between the borehole wall and the farfield changes gradually from large to small. Therefore, to reduce computer

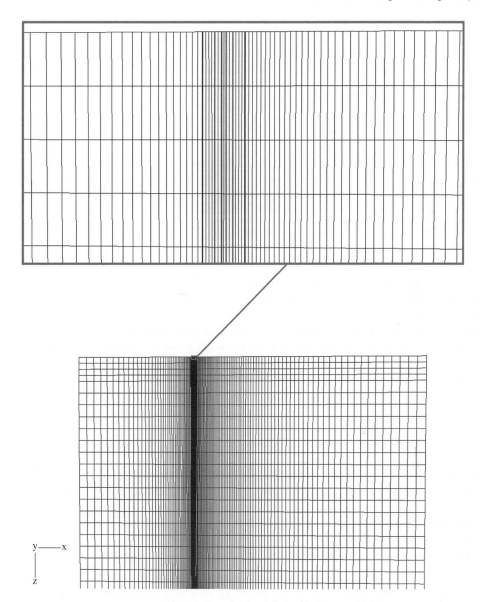

Figure 8.8 Computational grids used in the solution domain in *xz* cross section.

memory and computational time, the size of the mesh cells is chosen based on this gradual change. Applying all these techniques, a three-dimensional $15\,\mathrm{m} \times 15\,\mathrm{m} \times 60\,\mathrm{m}$ domain may require mesh sizes of the order of 1 000 000 elements to simulate multiple boreholes of 50 m length.

To define the control volumes in unstructured meshes, a cell-centered control volume technique is applied by ANSYS FLUENT. In the cell-centered method, the nodes are placed at the centroid of the control volume, as illustrated in Figure 8.9, while the boundary nodes reside at the center of boundary cell faces.

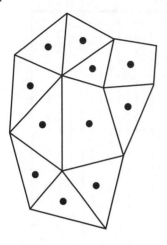

Figure 8.9 Cell-centered control volume construction in two-dimensional unstructured meshes.

8.2.8.2 Grid Quality

The quality of the mesh plays a significant role in the accuracy of estimating the temperature and stability of the numerical computation. The attributes associated with mesh quality are node point distribution, smoothness and skewness (non-orthogonality). Rapid changes in cell volume between adjacent cells translate into larger truncation errors. To improve the smoothness, the mesh should be refined based on the change in cell volume or the gradient of the cell volume. The central difference used in the discretization of the governing equations is only accurate if the mesh is fully orthogonal, that is when the line joining neighboring nodes P and A and the unit normal vector n_i are in the same direction.

Cell skewness is a non-dimensional parameter calculated using the normalized angle and volume deviation methods. It is a measure of the difference between the shape of the cell and the shape of an equilateral cell of equivalent volume or angle. Optimal quadrilateral triangular meshes have angles of close to 60° and have all angles less than 90°. Highly skewed cells can decrease accuracy and destabilize the solution. A cell skewness value of 0 indicates a best case equiangular and equilateral cell, and a value of 1 indicates a completely degenerate cell which is characterized by nodes that are nearly coplanar (collinear in two-dimensional).

In the current mesh, cell skewness is reported for the two-dimensional mesh as:

Maximum equivolume cell skewness = 0.53

Maximum equiangle cell skewness = 0.64

Moreover, cell squish is a measure used to quantify how far a cell deviates from orthogonality with respect to its faces. The worst cells have a Cell Squish Index close to 1. In the current mesh, cell squish is reported for the two-dimensional mesh as:

Maximum cell squish = 0.41

Maximum face squish = 0.28

To eliminate errors due to grid coarseness, a grid sensitivity study is performed. In general, the grid is made finer until less than 1% change in temperature rise calculated at the borehole wall is achieved.

8.2.9 Discretization

The discretization of unstructured meshes can be developed with a basic control volume technique where the integral form of the conservation equation for transport of a scalar quantity Φ is used as the starting point:

$$\int_{CV} \frac{\partial \rho \phi}{\partial t} dV + \oint \rho \phi \vec{v} \cdot d\vec{A} = \oint \Gamma_\phi \nabla \phi \cdot d\vec{A} + \int_{CV} S_\phi dV \tag{8.29}$$

where ρ is the density, $\vec{v}$ is the flow velocity vector, Γ_φ is the diffusion coefficient for φ, and S_φ is source of φ per unit volume. The terms on the left-hand side of Equation (8.29) are the conservative forms of the transient derivative of the transported variable φ and the convection terms, respectively, and the diffusion and source terms appear on the right-hand side. Discretization of Equation (8.29) yields the following general form:

$$\frac{\partial \rho \phi}{\partial t} V + \sum_{i}^{N_{faces}} \rho \phi_i \vec{v}_i \cdot \vec{A}_i = \sum_{i}^{N_{faces}} \Gamma_\phi \nabla \phi_i \cdot \vec{A}_i + S_\phi V \tag{8.30}$$

where ρ is the area of surface i, V is cell volume, and N_{faces} is the number of faces enclosing a cell, which depends on the cell topology. Face values of φ are required for the convection term in Equation (8.30) and must be interpolated from the cell center values using a second-order upwind scheme. The first-order upwind scheme is used when the flow is aligned with the mesh. In the current study, since the mesh is a triangular prism in shape, the flow crosses the mesh lines obliquely at the U-tube turn and first-order convective discretization may increase the numerical discretization error. To obtain more accurate results, second-order discretization is used. In this approach, quantities at cell faces are computed through a Taylor series expansion of the cell-centered solution about the cell centroid. Thus, when a second-order upwinding is selected, the face value φ_i is computed using the following expression:

$$\phi_i = \phi + \nabla \phi \cdot \vec{r} \tag{8.31}$$

where ϕ and $\nabla \phi$ are the cell-centered value and its gradient in the upstream cell, and $\vec{r}$ is the displacement vector from the upstream cell centroid to the face centroid. Equation (8.30) is non-linear with respect to the unknown scalar variable φ at the cell center as well as in surrounding neighbor cells (φ_{nb}). A linearized form of Equation (8.30) can be written as:

$$a_p \phi = \sum_{nb} a_{nb} \phi_{nb} + b \tag{8.32}$$

where a_p and a_{nb} are linearized coefficients for ϕ and ϕ_{nb}. Similar relations with the form of Equation (8.32) can be written for each cell in the solution domain, which leads to a system of algebraic equations. In ANSYS FLUENT, this system is solved via a Gauss–Seidel method (ANSYS n.d.).

8.2.10 Pressure-Based Solver

The current problem consists of three regions: ground; grout; and the borehole fluid. The ground and grout regions are solid regions and, therefore, the only mode of heat transfer is conduction. The borehole fluid region, however, employs all conservation equations: mass; momentum; and energy. In solving these equations, special practices are employed in the discretization of the continuity and momentum equation. Using the discretization scheme described in Section 8.2.9, the x-momentum equation can be obtained as:

$$a_p u = \sum_{nb} a_{nb} u_{nb} + \sum P_i \vec{A} \cdot \hat{i} + S \tag{8.33}$$

Equation (8.33) requires the value of the pressure P_i at the face between neighboring cells c_0 and c_1. If the pressure field is not known when solving Equation (8.33), an interpolation scheme is required to compute the face values of pressure from the cell values. Since, in the current example, the pressure variation between the cells is expected to be smooth, the pressure profile is not expected to have a high gradient at a cell face and, therefore, the following standard pressure interpolation scheme is used:

$$P_i = \frac{\dfrac{P_{c_0}}{a_{p,c_0}} + \dfrac{P_{c_1}}{a_{p,c_1}}}{\dfrac{1}{a_{p,c_0}} + \dfrac{1}{a_{p,c_1}}} \tag{8.34}$$

The continuity equation may be discretized as:

$$\sum_i^{N_{faces}} \rho v_i A_i = 0 \tag{8.35}$$

Here, it is necessary to relate the face values of velocity v_i to the stored values of velocity at the cell centers. ANSYS FLUENT uses momentum-weighted averaging, with weighting factors based on the a_p coefficient from Equation (8.33) to obtain the face flux ρv_i in Equation (8.35). Hence,

$$\rho v_i = \rho \hat{v}_i + d_f (p_{c_0} - p_{c_1}) \tag{8.36}$$

where $\hat{v}_i$ contains the influence of velocities in cells c_0 and c_1.

$$\hat{v}_i = \frac{a_{p,c_0} v_{n,c_0} + a_{p,c_1} v_{n,c_1}}{a_{p,c_0} + a_{p,c_1}} \tag{8.37}$$

Here, p_{c0}, p_{c1}, $v_{n,c0}$ and $v_{n,c1}$ are the pressures and normal velocities, respectively, within the two cells on either side of the face i. The term d_f is a function of $\bar{a}_p$, the average of the momentum equation a_p coefficients for the cells on either side of face i.

To couple pressure and velocity in the continuity and momentum equations, Equation (8.36) is used to derive an additional condition for pressure by reformatting the continuity equation [Equation (8.35)]. In the current study, the flow problem is solved in a pressure-based segregated manner by using the SIMPLE algorithm. This algorithm employs a relationship between velocity and pressure corrections to enforce mass conservation and to obtain the pressure field. In this algorithm, the momentum equation [Equation (8.33)] is solved using a guessed pressure field, p^*, resulting in a face flux ρv_i^* that is computed from Equation (8.36). Therefore,

$$\rho v_i^* = \rho \hat{v}_i^* + d_f (p_{c_0}^* - p_{c_1}^*) \tag{8.38}$$

If this face flux does not satisfy the continuity equation, a correction $\rho v_i'$ is added to the face flux ρv_i^*, so that the corrected heat flux is expressed:

$$\rho v_i = \rho v_i^* + \rho v_i' \tag{8.39}$$

and satisfies the continuity equation. In order to correct the initial guessed pressure, p^*, the SIMPLE algorithm assumes that the face flux correction, $\rho v'_i$, can be written as:

$$\rho v'_i = d_f(p'_{c_0} - p'_{c_1}) \tag{8.40}$$

where p' is a cell pressure correction. The flux correction equations [Equation (8.39) and Equation (8.40)] are then substituted into the discrete continuity equation [Equation (8.35)] to obtain a discrete equation for the pressure correction p' in the cell:

$$a_p p' = \sum_{nb} a_{nb} p'_{nb} + \sum_i^{N_{faces}} \rho v^*_i A_i \tag{8.41}$$

The second term on the right-hand side of Equation (8.41) is the net flow rate into the cell. The solution of Equation (8.41) can be used in correcting the cell pressure and the face flux:

$$p = p^* + \alpha_p p' \tag{8.42}$$

$$\rho v_i = \rho v^*_i + d_f(p'_{c_0} - p'_{c_1}) \tag{8.43}$$

where α_p is the under-relaxation factor for pressure. The corrected face flux ρv_i satisfies the discrete continuity equation identically during each iteration.

Since the current study involves transient operation of boreholes, the governing equations must be discretized in time in addition to space. For time-dependent flows, the pressure-based solver in ANSYS FLUENT uses an implicit discretization of the transport equation. Therefore, every term in the differential equations is integrated over time step Δt. In the pressure-based solver, the overall time-discretization error is determined from two sources of error: temporal discretization; and the manner in which the solutions are advanced to the next time step (time-advancement scheme).

A second-order implicit temporal discretization is used to replace the time integrals in the current study which appears in the following format:

$$\frac{3\phi^{n+1} - 4\phi^n + \phi^{n-1}}{2\Delta t} = F(\phi^{n+1}) \tag{8.44}$$

where the function F incorporates any spatial discretization. Note that ϕ^{n+1} is used in evaluating F due to the implicit method, that is, all convective, diffusive, and source terms are evaluated from the fields for time level $n+1$. Thus, ϕ^{n+1} cannot be expressed explicitly in terms of the existing solution values, ϕ_n. The implicit equation can be solved iteratively at each time level before moving to the next time step. The advantage of this scheme is that it is unconditionally stable with respect to time step size and introduces $O[(\Delta t)^2]$ truncation error.

The segregated solution process by which the equations are solved one by one introduces a splitting error. In ANSYS FLUENT, there are two approaches to the time-advancement scheme depending on how the splitting error is controlled: iterative time-advancement; and non-iterative time-advancement. The first scheme is chosen for time advancement in the current example since, in this scheme, non-linearity of the individual equations and inter-equation couplings are fully accounted for, eliminating the splitting error. All the equations are solved iteratively, for a given time step, until the convergence criteria are met. More specifically, the frozen flux formulation is used in the time advancement scheme. This formulation addresses the non-linear terms resulting from implicit discretization of the convective part of the transport equation and provides an optional way to discretize the convective part of the transport equation using the mass flux at the cell faces from the previous time level n. This reduces the non-linear character of the discretized transport equation without compromising accuracy and improves the convergence within each time step.

When including heat transfer in the borehole fluid region in the model, the pressure-based solver does not include the pressure work or kinetic energy when solving incompressible flow. Furthermore, in the current pressure-based solver, viscous dissipation terms, which describe the thermal energy created by viscous shear in the flow, are not included in the energy equation because viscous heating is negligible in the current problem. Details of the derivation of the energy equation in the ground and grout regions are given in Appendix A.

8.2.11 Initial and Boundary Conditions

The initial and boundary conditions used in the current model are included in Sections 8.2.3, 8.2.4, and 8.2.5, for the borehole fluid, grout and ground regions, respectively. In this section, additional information is provided on how such boundaries are set up and used in the iterations of the numerical solver. A uniform initial temperature of 282 K (equal to the undisturbed ground temperature) is assumed to be effective over the entire borefield. At the outer edge of the domain, a constant farfield temperature condition is applied ($T_0 = 282$ K).

The temperature and heat flux distributions on the borehole wall cannot be decided due to the dynamic nature of the heat exchange process between the pipes in the borehole and the borehole wall.

To fully account for the conjugated thermal process occurring in the borefield, the inlet temperature and flow rate of the fluid are specified as boundary conditions (see Section 8.2.3). The borehole inlet fluid temperature $T'_f(t)$ varies according to the temperature of the U-tube surrounding (grout) to maintain the required amount of heat flow rate during system operation.

To maintain the required ground heat load q', the inlet temperature of the borehole fluid varies as follows:

$$T'_f = \frac{q'H}{\dot{m}c_p} + T_{out} \tag{8.45}$$

In Equation (8.45), the borehole inlet fluid temperature T'_f needs to be updated at every time step for a constant heat flow rate per unit length since, as the system operates, the surrounding ground temperature rises/drops gradually resulting in a rise/drop in the borehole outlet fluid temperature T_{out}. If the average heat flow rate per unit length of borehole is transient, the borehole inlet fluid temperature changes according to the variation in outlet fluid temperature and the ground heat flow rate per unit length. The transient heat flow and borehole fluid temperature cannot be readily defined in ANSYS FLUENT. Therefore, in the current example, a user defined function is programmed in C and linked to the pressure-based solver. Details on the algorithm that is used in this code are given in Figure 8.10 and the code can be found elsewhere (Kouhi Fayegh Dehkordi 2013).

As mentioned in various sections throughout this book, the ground heat load depends on many parameters, including climate conditions, building type (residential,

Figure 8.10 Illustration of algorithm used in the user defined function applied in the simulation.

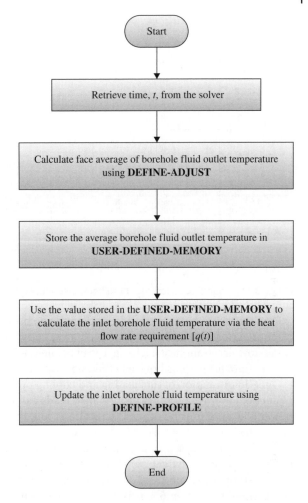

recreational, industrial, office, etc.) and size, number of people in the building, and GSHP specifications. The effect of each parameter on the heating and cooling profiles can be examined and the dominating factors on heating and cooling profile shape can be determined in the following form:

$$\dot{Q} = f(\dot{Q}_{\text{max}}, t) \tag{8.46}$$

Here, $\dot{Q}$ is the heating/cooling load at time t, Q_{max} is the maximum heating/cooling (which varies with the building size, number of building occupants and other factors), and f is a periodic function that best describes the shape of heating and cooling loads in a specified building type throughout the year. In order to examine the effects of such parameters on the building heating and cooling load profile shape, a heating and cooling load analysis can be performed using programs such as HvacLoadExplorer (McQuiston *et al.* 2005), and the design heating and cooling loads for each month based on the varying temperature profile can be obtained. Using these data, the form of the function f can be determined.

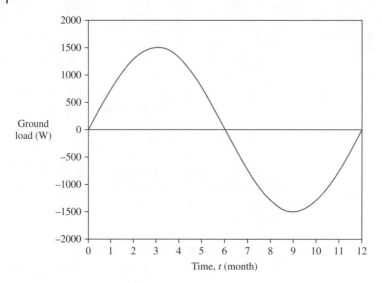

Figure 8.11 Ground load profile for both boreholes in the current modeling example.

Performance data for the heat pump unit [e.g., heating and cooling capacities, coefficient of performance (COP), electrical power consumption, water pressure drop at the water-to-refrigerant heat exchanger] can be modeled as functions of the entrance water temperature for various borehole fluid flow rates.

Once an estimation of the heating and cooling profile is determined in the form of an *f* function, average monthly ground heat loads can then be defined via the heat pump power consumption and other specifications, and then used as the heat transfer boundary on the borehole wall. In order to achieve a more accurate ground heat load profile, a typical heat pump system can be used and the ground heat load can be calculated experimentally or via analytical methods such as the bin method (McQuiston *et al.* 2005). A case study using bin method calculations to estimate the ground heat load profile is presented in Section 6.3. It is shown that, due to the periodic weather changes, the ground heat load profile is expected to be periodic. Here, it is assumed that the ground load profile is modeled as a simplified sinusoidal profile, as shown in Figure 8.11. More detailed analyses of parameters affecting heat exchanger operation can be found in studies reported in the literature focusing on such areas (Li and Lai 2013).

8.2.12   User Defined Function

To define a transient boundary condition in ANSYS FLUENT, a user defined function must be programmed in C and linked to the pressure-based solver. This is done here for the borehole inlet fluid temperature. The algorithm used in this code to evaluate the transient boundary condition at the borehole fluid inlet is illustrated in Figure 8.10. It is seen that, in the numerical approach, the temperature of the borehole fluid at the U-tube exit is calculated at every time step and a new temperature for the borehole fluid inlet temperature is set according to the required ground heat load for the next time step. Since, to maintain the required heat flow rate, the temperatures of the inlet and outlet borehole fluids must follow the heat flow rate occurring at the same time step, calculating the borehole inlet fluid temperature this way and using it as the boundary condition for the borehole inlet fluid temperature in the next time step can increase the error in the solution. This method is preferred due to its relative simplicity and reduced computation time in ANSYS FLUENT. However, in order to minimize the error in the results of the simulation, the time steps should be chosen small enough so that the change in the required heat flow rate and, consequently, the borehole inlet fluid temperature is kept relatively small from one time step to the next.

8.2.13 Summary

In the pressure-based approach, the pressure field is extracted by solving a pressure or pressure correction equation which is obtained from the continuity and momentum equations. ANSYS FLUENT solves the governing equations for the conservation of mass and momentum (when appropriate), and for energy. In both cases a control-volume-based technique is used that consists of:

- Division of the domain into discrete control volumes using a computational grid.
- Integration of the governing equations on the individual control volumes to construct algebraic equations for the discrete dependent variables ("unknowns'") such as temperature.
- Linearization of the discretized equations and solution of the resultant linear equation system to yield updated values of the dependent variables.

Each iteration consists of the steps illustrated in Figure 8.12 and summarized below:

- Update fluid properties (e.g., density, viscosity, specific heat) for the current solution.
- Allocate a user defined function for the borehole fluid inlet temperature.
- Update boundary conditions using a DEFINE-PROFILE macro.
- Solve the system of continuity, momentum and energy equations for the borehole fluid model.
- Solve the energy equations for the grout and ground models.
- Check for convergence of the equations.

After each irritation, the simulation time is checked and, if the simulation should continue, the temperature boundary condition at the borehole fluid inlet is updated according to the new borehole fluid outlet temperature and the required heat flow rate.

8.3   Numerical Modeling of a Borefield

When only the heat flow patterns outside the borehole are of interest (e.g., when examining environmental impacts of borehole systems on the surroundings), only the ground region outside the boreholes is modeled. Here, a borefield that includes 16 boreholes is modeled numerically to examine the temperature of the ground outside the borefield boundary. In this model, the building loads are calculated using a simplified bin method and real weather data (see Section 6.3). The transient nature of geothermal systems responding to the needs of buildings is described and challenges related to modeling such systems are discussed.

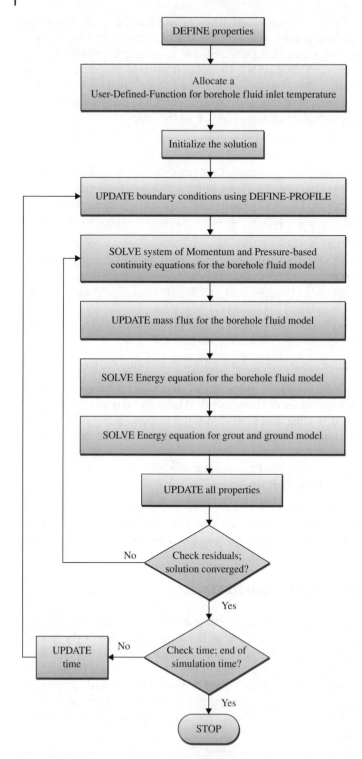

Figure 8.12 Solution procedure for modeling a domain consisting of multiple boreholes including the borehole fluid, grout, and ground in ANSYS FLUENT.

8.3.1 Physical Domain

Koohi-Fayegh and Rosen (2015) have shown that results of a two-dimensional model are valid for analysis of heat flow in the ground surrounding a borehole when a heat flux boundary condition is used at the borehole (see Appendix B, Section B.2). Therefore, a two-dimensional model of transient heat conduction in the ground is presented in this section. The domain consists of several borehole systems, each consisting of 16 vertical boreholes (Figure 8.13). The borehole systems are placed at every 100 m and the boreholes are installed at 6-m separation distances.

In the first stage of numerical analyses, a geometric model must be built. The heat transfer symmetry about the system shown in Figure 8.13a is utilized. Therefore, only a quarter of a borehole field is modeled and the solution domain (ground) is enclosed by the farfield and three symmetry planes. In Figure 8.13, the gray area is the solution domain, the results of which can be replicated to the other areas due to their symmetry. Here, the farfield representing the undisturbed ground is selected far enough from the boreholes to ensure the boundary temperature is maintained consistently at the value of the farfield temperature over the time of concern, that is, the amount of heat flux at the outer edge of the domain is zero or insignificantly small. The reason for selection of farfield rather than another symmetry for which thermal interaction can be examined is to determine the migration of the thermal plume to the undisturbed ground.

8.3.2 Boundary Conditions

For numerical heat transfer calculations, a uniform initial temperature of 282 K (equal to the undisturbed ground temperature) is assumed over the entire borefield. At the symmetry boundaries, there is no heat flux across the symmetry plane which results in zero normal gradients of temperature at the symmetry plane. At the outer edge of the domain, a constant farfield temperature condition equal to the initial temperature is applied (282 K) to obtain the closed-form solution to the heat transfer problem.

In evaluating the temperature rise in the ground due to the presence of GHEs, a key step is to define the heat flux from the surface of the heat exchanger to the ground. This

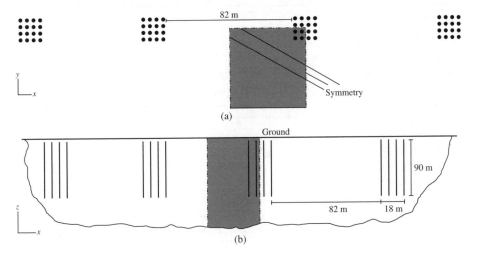

Figure 8.13 Solution domain. (a) Horizontal cross section (*xy*) and (b) horizontal cross section (*xz*).

can be complicated due to the dynamic nature of the heat transfer from the fluid flowing in the U-tubes within the borehole to the borehole wall. For simplicity, the U-tube configuration in the borehole is not simulated in the model and the boundary condition at the borehole wall is set to the heat flux. This is done since, when studying heat flow in the ground surrounding the boreholes and its possible negative effects on nearby ecosystems, their inner dynamic heat exchange processes can be of second priority compared with the heat dissipation in the ground surrounding them. Furthermore, it is assumed that the inlet temperature of the borehole fluid running in the U-tube inside the borehole is adjusted according to the building heating needs. The annual periodic variation in weather conditions is correlated with the heat flux at the wall of the heat exchanger (the borehole wall, here). In this example, the transient heat flux profile illustrated in Figure 6.11 is used as the heat flux boundary condition at the borehole wall. The heat flux profile is balanced, that is, the total amount of heat delivered to the ground in the cooling period is the same as the heat that is removed in the heating period. This assumption is made since, in the design stage of these systems, an objective is to have a balanced heat flow profile in order to avoid temperature rise or fall after each yearly cycle due to an unbalanced heat flux. Details on how such heat flux profiles can be determined are provided in Section 6.3. In order to account for the transient term in the conduction equation, the time is subdivided into time steps of 3600 s.

The geometric and thermal characteristics for the borehole and the surrounding ground are assumed to be as in Table 8.2.

Similar to the example in the previous section, GAMBIT is used to build the solution domain and define unstructured triangular cells as shown in Figure 8.14. The region nearest to the boreholes, where the temperature gradient is higher, is meshed more finely to enable the temperatures to be accurately predicted. The necessary parameters including the material thermal properties as well as the boundary conditions are defined

Table 8.2 Thermal properties and geometric characteristics used in the model.

Property	Value
Ground	
Undisturbed ground temperature	9°C (282 K)
Ground thermal conductivity	1.5 W/m K
Ground specific heat capacity	1200 J/kg K
Ground density	1381 kg/m^3
Borehole geometry	
Total borehole length, H	NA
Borehole radius, r_b	0.050 m
Number of boreholes	16
Borehole separation distance, D_b	6 m
Heat flow rate per unit length, q' (W/m)	Periodic [Equation (6.17) and Equation (6.18)]

NA, not applicable.
Source: Incropera and DeWitt, 2000; Gao *et al.* 2008; Hepbasli *et al.* 2003; Shonder and Beck 1999.

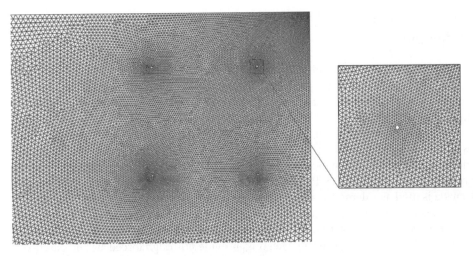

Figure 8.14 Triangular mesh used for the solution domain.

in ANSYS FLUENT. After the volume model is built in ANSYS FLUENT, the governing integral equations for heat conduction in the ground can be solved numerically.

A full numerical model has already been presented in Section 8.2 that solves the governing equations in the ground, grout and borehole fluid in a three-dimensional domain. For brevity, details on the pressure-based solver that can be used to complete this model are not provided here. However, since there is much similarity between the solver that can be used for the current model and the one used for the numerical model presented in Section 8.2, the reader is refered to Section 8.2.7 for more information on how such solvers are set up.

8.3.3 Model Limitations

One limitation in the current model is using a heat flux boundary condition on the borehole wall. As mentioned previously, it is assumed that the inlet temperature of the fluid circulating in the U-tube inside the borehole is adjusted according to the ground heat load. Since the borehole fluid temperature is one of the key parameters in the heat pump operation and efficiency, it is normally not adjusted to values that result in a low efficiency for the heat pump. Using a heat flux boundary condition can cause the temperature of the ground to rise infinitely without a stop in system operation. In reality, if the temperature of the ground surrounding a borehole becomes close to or higher than the inlet temperature of the borehole fluid exiting the heat pump, the system will not be able to deliver the desired heat to the ground and will automatically stop operating until the heat around it flows away and is dissipated and the temperature drops to a lower value. In order to overcome such a limitation, the periodic heat boundary on the borehole wall can be replaced with a temperature boundary condition, or a heat flux boundary condition that is related to the borehole fluid temperature and can be updated at short time steps with respect to the ground temperature. This is possible if the heat transfer model for outside the borehole is coupled to the model inside the borehole (see previous section).

8.4 Numerical Modeling of a Horizontal Ground Heat Exchanger

In this section, a complex model is provided of heat and moisture transfer in the ground for a system consisting of building heating/cooling loads, a heat pump unit, and heat transfer at a HGHE. The system's performance is analyzed using energy analysis. The impact of heat deposition and extraction on the underground environment in terms of temperature and moisture variations is also discribed. System design parameters, such as pipe length and depth and layout of a pipe loop, are considered in order to achieve the good thermal performance of the system and environmental sustainability of the ground. The design of a GSHP system for an existing residencial dwelling in Ontario, Canada is used as a demonstration of the simulation tool.

A two-storey old house located in Wasaga Beach, Ontario, Canada, with a total living floor area of 227 m^2 (2440 ft^2), including a finished basement, is considered. The house is equipped with electric resistance heating, provided through forced air and baseboard convection units. For space cooling, two window-mounted air conditioners are used for the most commonly used bedrooms.

8.4.1 Physical Domain

Two two-dimensional ground domains of rectangular shape are considered:

- a single layer in a series and serpentine layout
- a double layer in a series and serpentine layout

The HGHE occupies a ground region with depth of 6 m and width of half of their horizontal pipe spacing H (Figure 8.15).

The investigated depths of the single-layer GHE are 0.5, 1.0, 1.5, and 2.0 m (labeled Cases A, B, C, and D, respectively). For the double-layer configurations, the following combinations of pipe depths are considered: 0.5 and 1.0 m as Case AB, 0.5 and 1.5 m as Case AC, and 1.0 and 1.5 m as Case BC. For each case, GHE lengths ranging from 200 to 1000 m at 100-m increments are considered. In total, there are 63 simulations. A horizontal pipe spacing of $H = 0.5$ m is applied for all simulations (Figure 8.15).

All domain boundaries are assumed to be adiabatic and impermeable to moisture, except the ground surface at the position where energy and moisture balances are modeled. The two two-dimensional ground domains are discretized into triangular meshes using a computer code called Gradmesh (Thompson *et al.* 1999). The finest mesh sizes of about 25 mm are used around the pipes (GHE) and they expand radially from the pipe wall, with the coarsest meshes at the bottom of the ground domains. These mesh sizes are adequate to capture the highest temperature and moisture gradients around a GHE pipe (Piechowski 1999).

8.4.2 Numerical Solver

The simulations in this example are performed using Ground Heat Exchanger Analysis, Design and Simulation (GHEADS), a computer simulation tool for designing and analyzing GSHP systems. The computer package was initially developed by Tarnawski and Leong (1990) for GSHP systems operating with HGHEs. Specific details of GHEADS

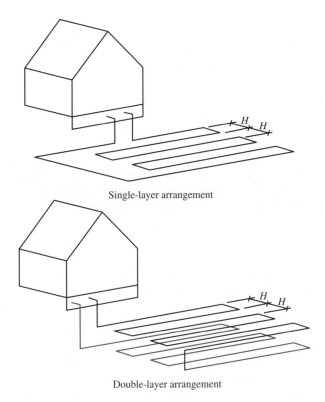

Single-layer arrangement

Double-layer arrangement

Figure 8.15 Horizontal ground heat exchanger arrangements.

have been reported by Tarnawski and Leong (1990, 1993) and Leong *et al.* (1998). The most important modeling approaches are reviewed in this section.

As previously mentioned in Section 7.1.2, moisture migration can significantly affect the operation of a HGHE. In order to solve the set of non-linear combined heat and moisture transfer equations representative of the heat and moisture flows in the ground surrounding HGHEs [Equation (7.2) and Equation (7.3)], the finite element method provides a number of advantages over the finite difference method. The advantages include the possibility of handling ground heat storage having complex geometries such as layers of different soils with different material properties and variable thicknesses, ground coil interfaces and ground surface topography.

GHEADS integrates several models to perform simulations of GSHP systems for any dwelling located at any location provided that the local climatic data and soil information are available. The moisture freezing/thawing process is modeled by an isothermal approach proposed by Hromadka *et al.* (1981). Time advancement of the solution is made using the three-time-level scheme proposed by Goodrich (1980). Energy and moisture balances at the ground surface follow the model used by Tarnawski (1982). The thermal and hydraulic properties of soils are evaluated by a special subroutine from a computer package called TheHyProS developed by Tarnawski and Wagner (1992, 1993). More information on these models is provided in Section 7.1.2 and by Leong *et al.* (2011).

To determine the GHE heat extraction from or injection to the ground, we use the log mean temperature difference (LMTD) method, which is iterative. By coupling an energy balance with a heat pump performance model, a converged solution is attained for the circulating fluid temperature entering the heat pump and the GHE heat transfer. The advantages of using the LMTD method are:

- It accounts reasonably well for variations of circulating fluid and soil temperatures along the GHE, provided that the fluid and soil temperatures at the GHE inlet and outlet and the overall heat transfer coefficient are determined properly.
- It reduces the three-dimensional ground domain along the GHE to two two-dimensional ground domains for solving purposes: domains at the GHE inlet and outlet planes. As the two two-dimensional domains are applicable for any length of GHE, a significant reduction is attained in computational time relative to the three-dimensional case.

Finally, for evaluating the heating and cooling loads of a dwelling throughout a year, a simple heating and cooling load model is employed that accounts for the most significant factors: effects of indoor–outdoor temperature difference; solar heat gain; and wind speed for infiltration. Alternatively, GHEADS can read a file of heating and cooling loads during the year.

8.4.3 Assumptions

In order to size a heat pump unit, a heating and cooling load analysis is made using a program called HvacLoadExplorer (McQuiston *et al.* 2005), and the design heating and cooling loads are obtained as 14.3 and 13.3 kW, respectively. Based on the design loads and climatic data, a simple load model is created for GHEADS to calculate heating and cooling loads. A commercially available water-to-air heat pump unit with a nominal 13.6 kW heating capacity is selected; it is based on heating capacity because of a much greater heating rather than cooling demand of the house. The performance data of the heat pump unit (such as heating and cooling capacities, COP, power consumption, water pressure drop at the water-to-refrigerant heat exchanger) are modeled as functions of the entrance water temperature (EWT). For a fluid flow rate of 53 l/min and an inlet air flow rate to the heat pump of 0.944 m^3/s, the entering air temperatures (EATs) of 21 and 24 °C are assumed for heating and cooling seasons, respectively. The heat pump is equipped with a single-speed scroll compressor and a thermostatic expansion valve for regulating refrigerant flow rate. When the heating or cooling capacity of the heat pump exceeds the heating or cooling load of the house, the GSHP system is operated under an ON–OFF condition.

In addition, the following factors are considered:

- The circulating fluid used in the GHE is a 20% (by weight) methanol antifreeze solution with a freezing point of about −12 °C. The heat pump unit has a lockout freeze protection for the circulating fluid set at −9.4 °C.
- Polyethylene pipe with a nominal size of 32 mm is used for the GHE. This size is selected to ensure turbulent flow in the pipe with low friction loss.
- The soil at the site is sand (bulk density of 1734 kg/m^3, porosity of 0.346, sand mass fraction of 0.852, silt mass fraction of 0.068, and clay mass fraction of 0.080).
- Climatic data for Ottawa, Ontario, Canada are used.

8.4.4   Performance Evaluation Method

Among all GHE configurations listed in Section 8.4.1, the best single-layer and double-layer cases are selected based on the highest overall thermal performance. For the best single-layer and double-layer cases, a wider horizontal pipe spacing of $H = 1.0$ m is tested in order to investigate the effectiveness of doubling the horizontal pipe spacing.

For a certain GHE length, the horizontal pipe spacing of $H = 1.0$ m requires twice as much ground surface area as needed with $H = 0.5$ m; but it is expected that the horizontal pipe spacing $H = 1.0$ m offers better performance because of the larger ground mass available for thermal energy storage.

Energy analysis is used to perform a thermodynamic performance comparison in the present example. The overall energy consumption of the GSHP system can be written as:

$$E_o = E_{HP} + E_{SUP} + E_{PUMP} \tag{8.47}$$

Here, E_{HP} is the total electrical energy supplied to the heat pump unit for operation of the unit including fan and compressor operation, E_{SUP} is the electric energy required for supplementary heating or cooling when the heat pump is unable to meet the demand, and E_{PUMP} is the total electrical energy supplied to a motor which drives a pump for circulating the antifreeze solution through the GHE and the water-to-refrigerant heat exchanger. It is assumed that the efficiencies of the motor and pump are 90 and 70%, respectively.

Also, for the GSHP system, we can write the seasonal heating COP as:

$$COP_H = \frac{Q_H}{(E_o)_H} \tag{8.48}$$

the seasonal cooling COP as:

$$COP_C = \frac{Q_H}{(E_o)_C} \tag{8.49}$$

and the overall COP as:

$$COP_o = \frac{Q_H + Q_C}{(E_o)_H + (E_o)_C} \tag{8.50}$$

Here, Q_H and Q_C are the heating and cooling provided by the heat pump, respectively.

8.4.5   Performance Analysis of the Ground-Source Heat Pump System

In this section, the performance of the GSHP system is analyzed based on single- and double-layer arrangements.

8.4.5.1   Single-Layer Arrangement

The seasonal COP of the GSHP system with a single-layer GHE at various depths and borehole lengths is examined for the current building.

8.4.5.1.1   Heating Season   Figure 8.16 shows the seasonal heating COP of the GSHP system with a single-layer GHE at various depths. The optimum pipe length, corresponding to highest COP_H, is 600 m, except for Cases B and D where the optimum pipe lengths are about 500 m.

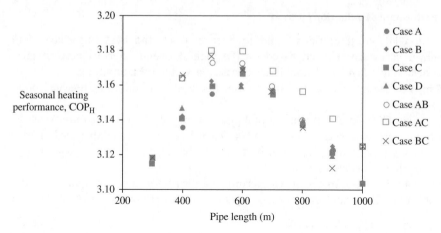

Figure 8.16 Variation of seasonal heating COP of the GSHP system with pipe length for a GHE at various depths.

When properly sized, Case A (0.5 m depth) has the highest COP_H of 3.17 among all the cases considered. But if the pipe length is shorter than the optimum length, the COP_H of Case A decreases drastically, for example, at a 200 m pipe length (not shown in Figure 8.16), COP_H becomes 2.5 and the temperature of the antifreeze solution drops to as low as −19 °C in the winter. However, under real operation, the heat pump would shut down earlier due to its lockout freeze protection.

At a pipe length of 1000 m, all single-layer GHE cases converge to the same COP_H of 3.1, as one would expect for an infinitely long pipe. The decline of COP_H when the pipe length is longer than the optimum is due to increased electrical power use for circulating an antifreeze solution. Essentially, COP_H is independent of GHE depth because the differences in COP_H, among the single-layer GHE cases at any pipe length, are negligibly small.

8.4.5.1.2 Cooling Season Figure 8.17 shows the seasonal cooling COP of the GSHP system with a single-layer GHE at various depths. The optimum pipe length corresponding to highest COP_C is approximately 600 m, except for Case A (0.5 m depth) whose highest COP_C is at a pipe length of about 900 m. The system performance depends more strongly on the depth of GHE.

A significant COP_C increase of 40.5% is observed at the optimum pipe length of 600 m for Case D with respect to Case A (Table 8.3). However, as the depth increases the increase of COP_C becomes less prominent, for example, COP_C increases by about 4.8% as the GHE depth changes from 1.5 to 2.0 m (Table 8.3).

Opposite to the heating operation, the best case for cooling is Case D (2.0 m depth) at the optimum pipe length. This is reasonable since deeper into the ground the temperature is cooler and more stable.

8.4.5.1.3 Overall Performance The overall (annual) performance of the GSHP system with a single-layer GHE is shown in Figure 8.18, and the COP_o values are relatively closer to the COP_H values than to the COP_C values. This is because the heating season is longer than the cooling season, causing it to require much more energy. The total heating and

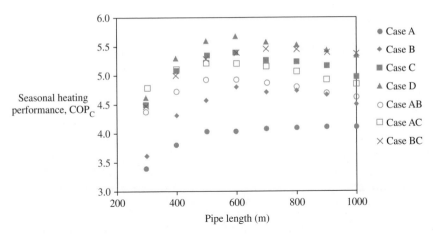

Figure 8.17 Variation of seasonal cooling COP of a GSHP system with pipe length for a GHE at various depths.

Table 8.3 Energy (MWh) and performance data for a single-layer GHE of 600 m at various depths.

Case (depth)	Q_H	Q_C	$(E_o)_H$	$(E_o)_C$	E_o	COP_H	COP_C	COP_o
A (0.5 m)	34.72	3.08	10.96	0.761	11.72	3.17	4.05	3.23
B (1.0 m)	34.73	3.11	10.99	0.647	11.64	3.16	4.81	3.25
C (1.5 m)	34.72	3.13	10.97	0.571	11.54	3.17	5.43	3.28
D (2.0 m)	34.66	3.13	10.97	0.556	11.53	3.16	5.69	3.28

cooling loads for the house are 30.63 and 2.79 MWh, respectively. Also the electrical energy consumption (e.g., of Case A) in the heating season is about 14 times more than that in the cooling season.

The peak COP_o is again at the optimum pipe length of 600 m, and both Cases C and D have the same COP_o of 3.28, even though Case C has a slightly greater overall energy consumption than Case D (11.54 vs 11.53 MWh, as indicated in Table 8.3). This difference is due to the fact that extra electrical energy consumption of 15 kWh in Case C returns 55 kWh more for heating and cooling (Table 8.3).

A similar situation exists for a GHE length of 500 m, as shown in Figure 8.19. The overall energy consumption for the pipe length of 500 m appears to be the lowest, but the extra energy consumption for a GHE length of 600 m returns more energy for heating and cooling. This is the reason why the thermal performance for a pipe length of 600 m is overall better than for a pipe length of 500 m.

In summary, the optimum pipe length for a single-layer GHE is 600 m, and it should be laid down at 1.5 m below the ground surface (Case C). For this design, 2 days in the winter require supplementary heating of 0.6 kW, for a total of 13 kWh; however, there is no requirement for supplementary cooling.

The annual energy requirement is 11.54 MWh. The yearly temperature variation of the antifreeze solution ranges from −6.5 to 27.6 °C, which seems to be environmentally acceptable.

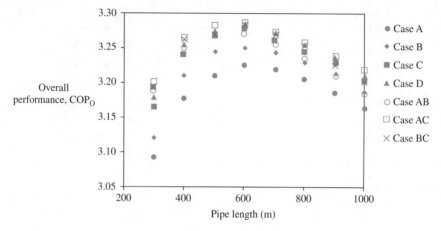

Figure 8.18 Variation of overall COP of GSHP system with pipe length for a GHE at various depths.

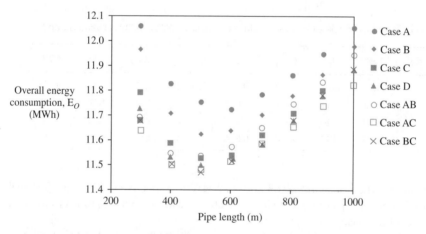

Figure 8.19 Variation of total energy consumption of GSHP system with pipe length for a GHE at various depths.

The required land area is 300 m² (for $H = 0.5$ m). Since the GHE at a 2.0 m depth does not offer notably better overall performance than the GHE at a 1.5 m depth, it is not considered for the simulations of double-layer GHEs.

8.4.5.2 Double-Layer Arrangement

In this section, the seasonal COP of the GSHP system with a double-layer GHE at various depths and borehole lengths is discribed for the current building.

8.4.5.2.1 Heating Operation Figure 8.16 shows the seasonal heating COP of the GSHP system with a double-layer GHE at various depths. The optimum pipe length appears to be around 500 m, which is 100 m shorter than the optimum pipe length for the single-layer GHE. Case AC has the highest COP_H, and it is also superior to all

Table 8.4 Energy (MWh) and performance data for a double-layer GHE of 500 m at various depths.

Case (depths)	Q_H	Q_C	$(E_o)_H$	$(E_o)o$	E_o	COP_H	COP_c	COP_c
AB (0.5/1.0 m)	34.59	3.11	10.91	0.628	11.54	3.17	4.94	3.27
AC (0.5/1.5 m)	34.59	3.11	10.89	0.593	11.48	3.18	5.20	3.28
BC (1.0/1.5 m)	34.56	3.11	10.89	0.584	11.47	3.18	5.31	3.28

single-layer GHE cases. Again, as for the single-layer GHE cases, the differences of COP_H among the double-layer GHE cases at any pipe length are negligible. This also indicates that COP_H is fairly independent of the depths of a double-layer GHE.

8.4.5.2.2 Cooling Operation Figure 8.17 shows the seasonal cooling COP of the GSHP system with a double-layer GHE at various depths. The optimum pipe lengths differ with depth (500 m for Case AB, 600 m for Case AC, and 800 m for Case BC).

It appears that the advantage of having a deeper buried GHE is beneficial for cooling operation. For example, for a fixed pipe length of 600 m, both Cases AB and AC exhibit a significant COP_C improvement relative to Case A.

8.4.5.2.3 Overall Performance The overall (yearly) performance of the GSHP system with a double-layer GHE is shown in Figure 8.18. Again, the COP_o values are closer to the ones for COP_H than to the ones for COP_C due to the same reason explained before. The peak COP_o occurs at the optimum pipe length of 500 m, and Cases AC and BC both have a COP_o of 3.28.

In summary, the optimum pipe length for a double-layer GHE is 500 m, and it should be laid down at 1.0 and 1.5 m below the ground surface (Case BC). For this design, 2 days in the winter require supplementary heating of 0.8 kW, for a total of 17 kWh; however, there is no requirement for supplementary cooling. The annual energy requirement is 11.47 MWh. The yearly temperature variation of antifreeze solution ranges from −6.9 to 31.6 °C, which also seems to be environmentally acceptable. The required land area is 125 m² (for $H = 0.5$ m). Comparing single-layer and double-layer GHEs in Figure 8.16, Figure 8.17, Figure 8.18 and Figure 8.19 and Table 8.3 and Table 8.4, the advantages of double-layer over single-layer GHEs can be seen.

8.4.5.3 Additional Simulations for $H = 1.0$ m
Two additional simulations are performed for Cases C and BC with a horizontal pipe spacing of $H = 1.0$ m (i.e., double the previous size). The results are tabulated in Table 8.5. Increasing horizontal pipe spacing improves the overall performance of both cases marginally. The improvement is slightly more significant for the single-layer GHE rather than for the double-layer GHE.

In fact, although the case of a single-layer GHE exhibits slightly better performance, it still requires about 62 kWh more electrical energy to operate than the double-layer GHE, mainly due to the additional pumping power for the extra 100 m pipe length. Moreover, the case of a single-layer GHE requires 350 m² more land area.

Table 8.5 Energy (MWh) and performance data for a doubled horizontal pipe spacing ($H = 1.0$ m).

Case (depth)	Pipe length (m)	Q_H	Q_C	$(E_o)_H$	$(E_o)_C$	E_o	COP_H	COP_C	COP_o
C (1.5 m)	600	34.84	3.12	10.87	0.635	11.50	3.21	4.93	3.30
BC (1.0/1.5 m)	500	34.59	3.11	10.83	0.613	11.44	3.20	5.04	3.29

8.4.6 Summary

The modeling and analysis of a GSHP system for heating and cooling applications are described and analyzed in this example. It is demonstrated that GHEADS is useful for simulating a GSHP system with a HGHE. The capabilities of the simulation tool are illustrated for designing an optimum GHE (configuration, length, and type) for an existing dwelling in Wasaga Beach, Ontario, Canada.

Many factors must be considered to optimize the design of a GHE. However, the optimum design and analysis are applicable only to a certain dwelling, because they are specific to the site characteristics (such as soil type and climatic conditions) and system operating parameters (such as magnitude and frequency of heating and/or cooling operations).

8.5 Model Comparison

As mentioned previously in this chapter, Sections 8.1 and 8.2 contain modeling approaches applied to the same problem: complete modeling of two neighbor boreholes. In this section, selected results of the analytical model presented in Section 8.1 are compared with those of the finite volume numerical method presented Section 8.2.

For the models presented in Sections 8.1 and 8.2, the borehole inlet fluid temperature is set to vary according to changes in the outlet fluid temperature and the transient ground heat flow rate per unit length. Figure 8.20 shows the change in the borehole wall temperature and borehole fluid inlet and outlet temperatures for the sinusoidal heat load profile in Figure 8.11. It is seen that, as the system experiences a periodic profile of ground heating and cooling load (Figure 8.11), the temperature of the borehole also experiences a periodic profile with a time lag. The maximum temperature of the borehole wall occurs some time after maximum heat input in the ground. This is due to the thermal capacitance of the borehole grout that results in a slower response to the change in its thermal condition. In the analytical method, the change in borehole wall temperature with time sets a new temperature for the borehole fluid inlet and outlet temperatures in order to deliver/remove a required heat load to/from the ground. It is seen that the difference between the borehole fluid inlet and outlet temperatures varies over time. This is due to the variable ground heat profile of the borehole. A comparison of the analytical and the numerical results for the borehole wall temperature and the borehole fluid inlet and outlet temperatures shows errors of less than 1%, where error is calculated as follows:

$$\text{Error} = \frac{T_{\text{numerical}} - T_{\text{analytical}}}{T_{\text{numerical}}} \tag{8.51}$$

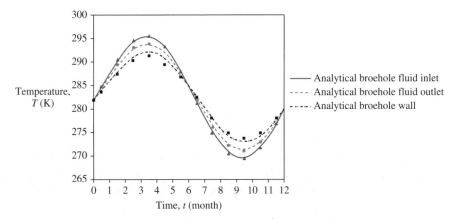

Figure 8.20 Transient temperature of the borehole wall and borehole inlet and outlet fluid temperatures for numerical and analytical solutions.

The error detected in the results of borehole wall temperature could be due to the simplifying assumptions in the analytical approach such as heat transfer from a line source of heat to the surrounding ground. With this assumption, the thermal properties of grout in the borehole are not accounted for when calculating the borehole wall temperature. Instead, to estimate the borehole wall temperature at $R = 0.1$ m, it is assumed that the line source of heat is surrounded by ground (neglecting the grout region) from $R = 0$ up to $R = R_b = 0.1$ m. Due to the small errors involved in the estimation of the borehole wall temperature and the inlet and outlet borehole fluid temperatures, it can be inferred that the analytical approach is capable of coupling the ground temperature at the borehole wall to the borehole fluid temperature when the ground load is known.

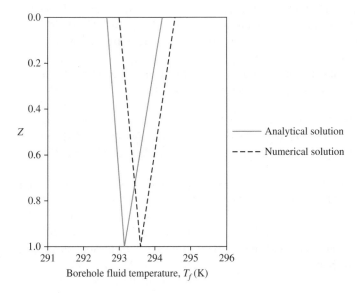

Figure 8.21 Borehole fluid temperature profile along the borehole from numerical and analytical solutions. Z is defined in Equation (7.31).

With the numerical approach, the temperature of the borehole fluid at the exit is calculated at every time step and a new temperature for the borehole fluid inlet temperature is set according to the required ground heat load for the next time step. In this approach, since the temperature of the inlet fluid is updated one time step after the outlet fluid temperature is calculated, the time steps should be chosen to be small enough that the change in the borehole inlet fluid temperature remains relatively small.

The borehole fluid temperatures of the two analytical and numerical models after 2.5 months of system operation are compared in Figure 8.21. It is seen that the temperature of the borehole fluid is 294.5 K at its inlet and, through losing heat to the surrounding ground, its temperature decreases along the U-tube length and drops to about 293.0 K at the outlet. A comparison of the temperatures resulting from both numerical and analytical models shows good agreement (less than 0.5 K difference). Therefore, both models are believed to be able to estimate the temperature of the borehole fluid along the borehole length. Note that the largest differences in the temperatures from the numerical and analytical models occur at higher heat loads in Months 2–4 and 8–10 (Figure 8.11), while the maximum 0.5 K temperature difference between the numerical and analytical results occurs for the entire system operation period.

References

ANSYS (n.d) ANSYS FLUENT 12.0 Theory Guide. http://www.afs.enea.it/project/neptunius/docs/fluent/html/th/node359.htm (accessed January 29, 2016).

Bauer, D., Heidemann, D. and Diersch, H.-J.G. (2011) Transient 3D analysis of borehole heat exchanger modeling. *Geothermics*, **40**, 250–260.

Bernier, M., Pinel, A., Labib, P. and Paillot, R. (2004) A multiple load aggregation algorithm for annual hourly simulations of GCHP systems. *HVAC&R Research*, **10** (4), 471–487.

Gao, J., Zhang, X., Liu, J. *et al.* (2008) Numerical and experimental assessment of thermal performance of vertical energy piles: An application. *Applied Energy*, **85**, 901–910.

Goodrich, L.E. (1980) Three-time-level methods for the numerical solution of freezing problems. *Cold Regions Science and Technology*, **3**, 237–242.

Hepbasli, A., Akdemir, O. and Hancioglu, E. (2003) Experimental study of a closed loop vertical ground source heat pump system. *Energy Conversion and Management*, **44**, 527–548.

Hromadka, T.V. II,, Guymon, G.L. and Berg, R.L. (1981) Some approaches to modelling phase change in freezing soils. *Cold Regions Science and Technology*, **4**, 137–145.

Incropera, F.P. and DeWitt, D.P. (2000) *Introduction to Heat Transfer*, 3rd edn, John Wiley & Sons, Inc., New York.

Koohi-Fayegh, S. and Rosen, M.A. (2015) Three dimensional analysis of thermal interaction of multiple vertical ground heat exchangers. *International Journal of Green Energy*, **12** (11), 1144–1150.

Kouhi Fayegh Dehkordi, S 2013 Thermal sustainability of low-temperature geothermal energy systems: system interactions and environmental impacts. PhD thesis. University of Ontario Institute of Technology.

Leong, W.H., Tarnawski, V.R. and Aittomäki, A. (1998) Effect of soil type and moisture content on ground heat pump performance. *International Journal of Refrigeration*, **21**, 595–606.

Leong, W.H., Tarnawski, V.R., Koohi-Fayegh, S. and Rosen, M.A. (2011) Ground thermal energy storage for building heating and cooling, in *Energy Storage* (ed. M.A. Rosen), Nova Science Publishers, New York, pp. 421–440.

Li, M. and Lai, A.C.K. (2013) Thermodynamic optimization of ground heat exchangers with single U-tube by entropy generation minimization method. *Energy Conversion and Management*, **65**, 133–139.

Marcotte, D. and Pasquier, P. (2008) Fast fluid and ground temperature computation for geothermal ground-loop heat exchanger systems. *Geothermics*, **37**, 651–665.

McQuiston, F.C., Parker, J.D. and Spitler, J.D. (2005) *Heating, Ventilating, and Air Conditioning: Analysis and Design*, 6th edn, John Wiley & Sons, Inc., Hoboken.

Piechowski, M. (1999) Heat and mass transfer model of a ground heat exchanger: Theoretical development. *International Journal of Energy Research*, **23**, 571–588.

Shonder, J.A. and Beck, J.V. (1999) Determining effective soil formation thermal properties from field data using a parameter estimation technique. *ASHRAE Transactions*, **105**, 458–466.

Tarnawski, VR 1982 An Analysis of Heat and Moisture Movement in Soils in the Vicinity of Ground Heat Collectors for Use in Heat Pump Systems', Acta Polytechnica Scandinavica, Mechanical Engineering Series 82, Helsinki.

Tarnawski, V.R. and Leong, W.H. (1990) *Computer Simulation of Ground Coupled Heat Pump Systems*, Technical Report, National Research Council, Ottawa.

Tarnawski, V.R. and Leong, W.H. (1993) Computer analysis, design and simulation of horizontal ground heat exchangers. *International Journal of Energy Research*, **17**, 467–477.

Tarnawski, V.R. and Wagner, B. (1992) A new computerized approach to estimating the thermal properties of unfrozen soils. *Canadian Geotechnical Journal*, **29**, 714–720.

Tarnawski, V.R. and Wagner, B. (1993) Modeling the thermal conductivity of frozen soils. *Cold Regions Science and Technology*, **22**, 19–31.

Thompson, J.F., Soni, B.K. and Weatherill, N.P. (1999) *Handbook of Grid Generation*, CRC Press, Boca Raton, pp. A.16–A.17.

Versteeg, H.K. and Malalasekera, W. (2007) *An Introduction to Computational Fluid Dynamics, The Finite Volume Method*, Prentice Hall, Harlow.

Yavuzturk, C. and Spitler, J. (1999) A short time step response factor model for vertical ground loop heat exchangers. *ASHRAE Transactions*, **105**, 475–485.

Zeng, H.Y., Diao, N.R. and Fang, Z. (2002) A finite line-source model for boreholes in geothermal heat exchangers. *Heat Transfer Asian Research*, **31** (7), 558–567.

Zeng, H.Y., Diao, N.R. and Fang, Z. (2003) Heat transfer analysis of boreholes in vertical ground heat exchangers. *International Journal of Heat and Mass Transfer*, **46** (23), 4467–4481.

9

Thermodynamic Analysis

9.1 Introduction

Although the use of the exergy concept is sometimes considered a new method for analyzing energy systems, its underlying fundamentals were introduced long ago with the formulation of the second law of thermodynamics. Section 2.2 included general aspects of the thermodynamic analysis and exergy method. In this chapter the application of thermodynamics and the exergy concept in analysis of energy systems to evaluate and optimize their design and performance is described. The energy systems discussed here are an underground thermal storage in the form of an aquifer thermal energy storage (TES), a hybrid ground-source heat pump (GSHP) system, and GSHPs and underground thermal storage.

9.2 Analysis of an Underground Thermal Energy Storage System

Energy and exergy analyses are presented for an underground thermal storage in the form of an aquifer TES in this section. Data are used from the first of four short-term aquifer TES test cycles, using the Upper Cambrian Franconia-Ironton-Galesville confined aquifer in Minnesota. The storage and supply wells are located 255 m apart. The test cycles were performed at the St. Paul campus of the University of Minnesota in the time frame November 1982–December 1983, during which time the average ambient temperature was reported to be 11 °C (Hoyer *et al.* 1985), and the analyses closely follow previous treatments (Rosen 1999; Dincer and Rosen 2011).

The tests involved three primary operating periods, during which the water temperature (Figure 9.1) and volumetric flow rate (Figure 9.2) vary temporally:

- **Charging.** Water is pumped from the source well, heated in a heat exchanger and returned to the aquifer through the storage well. The charging period occurred during 5.24 days over a 17-day period. The charging temperature and volumetric flow rate were approximately constant at mean values of 89.4 °C and 18.4 l/s, respectively.
- **Storing.** The storage period lasted 13 days.
- **Discharging.** After storage, energy was recovered by pumping the stored water through a heat exchanger and returning it to the supply well. Discharging occurred over 5.24 days, with a nearly constant volumetric flow rate (with a mean value of

Geothermal Energy: Sustainable Heating and Cooling Using the Ground, First Edition.
Marc A. Rosen and Seama Koohi-Fayegh.
© 2017 John Wiley & Sons, Ltd. Published 2017 by John Wiley & Sons, Ltd.

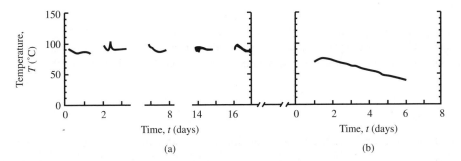

Figure 9.1 Temporal variations during experimental aquifer thermal energy storage test cycles of water temperature during charging (a) and discharging (b).

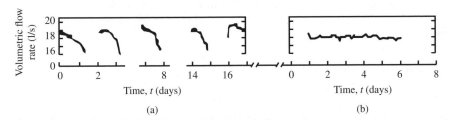

Figure 9.2 Temporal variations during experimental aquifer thermal energy storage test cycles of water volumetric flow rate during charging (a) and discharging (b).

18.1 l/s) and a linearly decreasing temperature with time (starting at 77 °C and ending at 38 °C).

9.2.1 Energy and Exergy Analyses

To facilitate the analyses, we treat the system in the simplified manner described in this subsection. During charging, heated water at a constant temperature T_c is injected at a constant mass flow rate $\dot{m}_c$ into the aquifer over a period t_c. After a storing period, discharging occurs over a period t_d, during which water is extracted from the aquifer at a constant mass flow rate $\dot{m}_d$. The fluid discharge temperature is a function of time, that is, $T_d = T_d(t)$, with the discharge temperature after an infinite time being the temperature of the reference- environment T_o, that is, $T_d(\infty) = T_o$, and the initial discharge temperature being between the charging temperature and the reference-environment temperatures, that is, $T_o \leq T_d(0) \leq T_c$. For simplicity, a discharge temperature–time profile is considered in which the discharge temperature decreases linearly with time from an initial value $T_d(0)$ to a final value T_o. The final temperature is reached at a time t_f and remains fixed at T_o for all subsequent times:

$$T_d(t) = \begin{cases} T_d(0) - (T_d(0) - T_o t/t_f, & 0 \leq t \leq t_f \\ T_o, & t_f \leq t \leq \infty \end{cases} \tag{9.1}$$

Throughout, pump work is neglected, and the working fluid, water, is assumed incompressible and taken to have a constant specific heat c. The temperature of the aquifer and its surroundings prior to heat injection is the reference-environment temperature T_o and only heat storage above T_o is considered.

9.2.1.1 Energy and Exergy Flows During Charging and Discharging

The energy and exergy injected during charging and recovered during discharging are evaluated. To determine these quantities, we note that the energy flow associated with a flow of liquid at a constant mass flow rate $\dot{m}$ for an arbitrary period of time with T a function of t, is:

$$E = \int_t \dot{E}(t)dt \tag{9.2}$$

where the integration is performed over the time period, and the energy flow rate at time t is:

$$\dot{E}(t) = \dot{m}c[T(t) - T_o] \tag{9.3}$$

and c is as previously defined. Also, the corresponding exergy flow is:

$$Ex = \int_t Ex(t)dt \tag{9.4}$$

where the exergy flow rate at time t is:

$$\dot{Ex}(t) = \dot{m}c\{[T(t) - T_o] - T_o \ln[T\{t\}/T_o]\} \tag{9.5}$$

For constant $\dot{m}$, c and T_o, we obtain the energy flow by combining Equation (9.2) and Equation (9.3):

$$E = \dot{m}c \int_t (T(t) - T_o)dt \tag{9.6}$$

and the exergy flow by combining Equation (9.4) and Equation (9.5), and substituting Equation (9.6):

$$Ex = \dot{m}c \int_t [(T(t) - T_o) - T_o \ln(T(t)/T_o)]dt = E - \dot{m}cT_o \int_t \ln(T(t)/T_o)dt \tag{9.7}$$

During charging, the energy and exergy input to the aquifer TES, for a constant water injection rate $\dot{m}_c$ and over a time period beginning at zero and ending at t_c, are expressed by Equation (9.6) and Equation (9.7), respectively, with $T(t) = T_c$:

$$E_c = \dot{m}_c c \int_{t=0}^{t_c} (T_c - T_o)dt = \dot{m}_c ct_c(T_c - T_o) \tag{9.8}$$

$$Ex_c = \dot{m}_c ct_c[(T_c - T_o) - T_o \ln(T_c/T_o)] = E_c - \dot{m}_c ct_c T_o \ln(T_c/T_o) \tag{9.9}$$

During discharging, the energy recovered from the aquifer TES, for a constant water recovery rate $\dot{m}_d$ and for a time period starting at zero and ending at t_d, is expressed by Equation (9.6) with $T(t)$ from Equation (9.1):

$$E_d = \dot{m}_d c \int_{t=0}^{t_d} [T_d(t) - T_o] dt = \dot{m}_d c[T_d(0) - T_o]\theta(2t_f - \theta)/(2t_f) \tag{9.10}$$

where

$$\theta = \begin{cases} t_d, & 0 \leq t_d \leq t_f \\ t_f, & t_f \leq t_d \leq \infty \end{cases} \tag{9.11}$$

The corresponding exergy recovery during discharging is expressed by Equation (9.7), with the same conditions as for E_d:

$$Ex_d = \dot{m}_d c \int_{t=0}^{t_d} [(T_d(t) - T_o) - T_o \ln(T_d(t)/T_o)]dt$$

$$= E_d - \dot{m}_d c T_o \int_{t=0}^{t_d} \ln(T_d(t)/T_o)dt \tag{9.12}$$

Here,

$$\int_{t=0}^{t_d} \ln[T_d(t)/T_o]dt = \int_{t=0}^{t_d} \ln(at + b)dt = [(a\theta + b)/a] \ln(a\theta + b) - \theta - (b/a)\ln b \tag{9.13}$$

where

$$a = [T_o - T_d(0)]/(T_o t_f) \tag{9.14}$$

$$b = T_d(0)/T_o \tag{9.15}$$

The expression for the integral in Equation (9.13) reduces when $t_d \geq t_f$ to:

$$\int_{t=0}^{t_d} \ln[T_d(t)/T_o]dt = t_f \left[\frac{T_d(0)}{T_d(0) - T_o} \ln \frac{T_d(0)}{T_o} - 1 \right] \tag{9.16}$$

9.2.1.2 Energy and Exergy Balances, Efficiencies and Losses

An aquifer TES energy balance taken over a complete charging–discharging cycle indicates that the input energy is either recovered or lost, while an exergy balance indicates that the input exergy is either recovered or lost, but lost exergy is associated with both waste exergy emissions and internal exergy consumptions due to irreversibilities.

Defining f is as the fraction of input energy E_c that can be recovered if the length of the discharge period approaches infinity (i.e., water is extracted until all recoverable energy has been recovered), we can write:

$$E_d(t_d \rightarrow \infty) = fE_c \tag{9.17}$$

An energy balance thus indicates that $(1 - f)E_c$ is the energy irreversibly lost from the aquifer TES, and that f varies between one for a storage without energy losses during an infinite discharge period to zero for a thermodynamically valueless storage. Exergy losses can occur even if $f = 1$, because the aquifer TES remains subject to mixing losses that reduce the temperature of the recovered water. Since E_c is given by Equation (9.8) and $E_d(t_d \rightarrow \infty)$ by Equation (9.10) with $\theta = t_f$, Equation (9.17) can be rewritten:

$$f = \frac{t_f \dot{m}_d [T_d(0) - T_o]}{2t_c \dot{m}_c (T_c - T_o)} \tag{9.18}$$

Since $T_d(0)$ can vary from T_o to T_c, the temperature-related term $[T_d(0) - T_o]/(T_c - T_o)$ varies between unity and zero. Both the time ratio t_f/t_c and the mass-flow-rate ratio $\dot{m}_d/\dot{m}_c$ can take on positive values, subject to the above equality.

The energy (or exergy) efficiency of the aquifer TES is defined as the fraction over a complete cycle of the energy (or exergy) input during charging that is recovered during discharging. As a function of the discharge time period, the energy efficiency η and the exergy efficiency ψ can be expressed, respectively, as:

$$\eta(t_d) = \frac{E_d(t_d)}{E_c} = \frac{\dot{m}_d[T_d(0) - T_o]}{\dot{m}_c(T_c - T_o)}\frac{\theta(2t_f - \theta)}{2t_f t_c} \tag{9.19}$$

$$\psi(t_d) = Ex_d(t_d)/Ex_c \tag{9.20}$$

The energy efficiency simplifies when the discharge period t_d exceeds t_f as follows:

$$\eta(t_d \geq t_f) = f \tag{9.21}$$

Thus, for an aquifer TES in which all injected energy is recoverable during an infinite discharge period, that is, $f = 1$, the energy efficiency can reach 100% if the discharge period t_d is sufficiently long. The corresponding exergy efficiency, however, remains less than 100% due to mixing losses, which cause much of the heat to be recovered at near-environmental temperatures.

The energy (or exergy) loss of the aquifer TES is defined over a complete cycle as the difference between input and recovered energy (or exergy). As a function of the discharge time period, therefore, the energy loss for the aquifer TES system can be expressed by $E_c - E_d(t_d)$ and the exergy loss by $Ex_c - Ex_d(t_d)$. The two main contributors to the thermodynamic losses are as follows:

- **Mixing.** As heated water is pumped into an aquifer TES, it mixes with the (usually cooler) water present, reducing the recovered water temperature to lower than that of the injected water. This loss renders the discharge temperature T_d at all times less than or equal to the charging temperature T_c, but not below the reference-environment temperature T_o [i.e., $T_o \leq T_d(t) \leq T_c$ for $0 \leq t \leq \infty$]. Exergy losses reflect the temperature degradation associated with mixing, while energy losses do not.
- **Unrecoverable heat transfer.** Energy input to an aquifer TES that is not recovered is lost. Thus, energy losses include energy remaining in the aquifer at a point where it could still be recovered if pumping were continued, and energy injected into the storage that is convected in a water flow or is transferred by conduction or other mechanisms sufficiently far from the discharge point that it is unrecoverable, regardless of how much or how long water is pumped out of the aquifer TES. Non-zero energy losses imply less than 100% of the injected energy is recoverable after storage.

Note that a threshold temperature T_t is usually applied, below which it is not worth continuing discharging since the residual energy in the aquifer water is of low quality. For the linear temperature–time relation here [Equation (9.1)], the appropriate discharge period can be evaluated using Equation (9.1) with T_t replacing $T_d(t)$ for the case where $T_o \leq T_t \leq T_d(0)$. Thus,

$$t_d = \begin{cases} \dfrac{T_d(0) - T_t}{T_d(0) - T_o} t_f, & T_o \leq T_t \leq T_d(0) \\[2ex] 0, & T_d(0) \leq T \end{cases} \tag{9.22}$$

Although thermal energy cannot be recovered if the threshold temperature exceeds the initial discharge temperature, the threshold temperature is nonetheless important because, as the discharge period increases, water is recovered from an aquifer TES at ever decreasing temperatures (ultimately approaching the reference-environment temperature), and the energy in the recovered water is of decreasing quality. Exergy analysis reflects this effect, as the recovered exergy decreases as the recovery temperature declines.

9.2.2 Assumptions and Simplifications

Several assumptions and simplifications are invoked:

- Mean values for volumetric flow rates and charging temperature are considered.
- The following water properties are considered constant: specific heat (4.2 kJ/kg K) and density (1000 kg/m^3).
- The reference-environment temperature T_o is taken to be constant at the mean ambient temperature of 11 °C.

9.2.3 Results and Discussion

The mass flow rates during charging and discharging are $\dot{m}_c = 18.4$ kg/s and $\dot{m}_d = 18.1$ kg/s, respectively, since the volumetric flow rate (in l/s) is equal to the mass flow rate (in kg/s) when the density is 1000 kg/m^3.

9.2.3.1 Energy and Exergy Flows During Charging and Discharging

During charging, it can be shown using Equation (9.8) and Equation (9.9), with $t_c = 5.24$ days (453 000 s) and $T_c = 89.4$ °C (362.4 K), that

$$E_c = (18.4\,\text{kg/s})(4.2\,\text{kJ/kg K})(453,000\,\text{s})(89.4°\text{C} - 11°\text{C}) = 2740\,\text{GJ}$$

and

$$Ex_c = 2740\,\text{GJ} - (18.4\,\text{kg/s})(4.2\,\text{kJ/kg K})\,(453,000\,\text{s})\,(284\,\text{K})\,\ln(362.4\,\text{K}/284\,\text{K})$$
$$= 320\,\text{GJ}$$

During discharging, the value of the time t_f is evaluated using the linear temperature–time relation of the present model and the observations that $T_d(t = 5.24\,\text{days}) = 38°\text{C}$ and $T_d(0) = 77$ °C (350 K). Then, using Equation (9.1) with $t = 5.24$ days,

$$38°\text{C} = 77°\text{C} - (77°\text{C} - 11°\text{C})(5.24\,\text{d}/t_f)$$

On solving, we find the final temperature is reached at $t_f = 8.9$ days, which is the discharge period when the discharge water temperature reaches T_o. Note that the rate of temperature decline would likely decrease in reality, causing the discharge temperature to asymptotically approach T_o.

The value of the fraction f, which represents the maximum energy efficiency achievable, can be evaluated with Equation (9.18) as:

$$f = \frac{(8.87\,\text{d})(18.1\,\text{kg/s})(77°\text{C} - 11°\text{C})}{2(5.24\ \text{d})(18.4\ \text{kg/s})(89.4°\text{C} - 11°\text{C})} = 0.701$$

For these values, it can be shown with Equation (9.14) and Equation (9.15) that

$$a = (11°\text{C} - 77°\text{C})/(284\,\text{K} \times 8.87\,\text{d}) = -0.0262\,\text{d}^{-1}$$

Table 9.1 Temporal variations for aquifer thermal energy storage, during discharging, of charged, recovered and lost energy and exergy as well as energy and exergy efficiencies.

Quantity	Discharge time, t_d (days)					
	0.0	2.2	4.4	6.7	8.9 ($= t_f$)[a]	>8.9
Energy quantities						
Charged (GJ)	2740					
Loss (GJ)	2740	1870	1280	920	810	810
Recovered (GJ)	0	810	1400	1750	1860	1860
Efficiency (%)	0	30	51	63	70	70
Exergy quantities						
Charged (GJ)	320					
Loss (GJ)	320	240	210	200	180	180
Recovered (GJ)	0	80	110	120	130	130
Efficiency (%)	0	24	35	39	40	40

a)   The time the final temperature is reached t_f in the case study is 8.9 days.

and

$$b = (350\,\mathrm{K})/(284\ \mathrm{K}) = 1.232$$

Values for charge energy E_c, charge exergy Ex_c, discharge energy E_d and discharge exergy Ex_d are presented as a function of discharge time period t_d (Table 9.1). The charged energy and exergy are only shown for a discharge time of zero, since the charging period is over before discharging begins. Note that for times greater than the time of 8.9 days required to reach the final temperature (last column in Table 9.1), the values are the same for those at a time of 8.9 days.

A normalized version of Table 9.1 is presented in Table 9.2. There, the time scale is normalized to show the time ratio t_d/t_f. Also, the charge energy, discharge energy and energy loss are presented as a percentage of the charge energy. Similarly, the charge exergy, discharge exergy and exergy loss are presented as a percentage of the charge exergy.

9.2.3.2   Energy and Exergy Efficiencies and Losses

The energy loss ($E_c - E_d$) and exergy loss ($Ex_c - Ex_d$) as well as the discharge energy efficiency η and discharge exergy efficiency ψ are given as a function of discharge time period t_d in Table 9.1. As t_d rises, the energy and exergy efficiencies in Table 9.1 both increase from zero to maximum values, and the difference between the two efficiencies increases. Thus, the exergy efficiency weights the energy recovered at higher t_d values less than the energy efficiency, since that energy is recovered at a temperature nearer to that of the reference environment. Note in Table 9.1 that as t_d approaches t_f (1) all parameters level off for the conditions specified and remain constant for $t_d \geq t_f$, while (2) the energy recovered increases from zero to a maximum value and the energy loss decreases from a maximum of all the input energy to a minimum (but non-zero) value. The exergy recovery and exergy loss functions exhibit much lower magnitudes but similar profiles.

Table 9.2 Normalized temporal variations for aquifer thermal energy storage, during discharging, of charged, recovered and lost energy and exergy

Quantity	Ratio of discharge time to time final temperature is reached, t_d/t_f [a]					
	0.00	0.25	0.50	0.75	1.00	>1.00
Energy quantities						
Charged percentage[b]	100					
Loss percentage	100	70	49	37	30	30
Recovered percentage (or efficiency)	0	30	51	63	70	70
Exergy quantities						
Charged percentage[c]	100					
Loss percentage	100	76	65	61	60	60
Recovered percentage (or efficiency)	0	24	35	39	40	40

a) The time the final temperature is reached t_f in the case study is 8.9 days.
b) The charged energy is 2740 GJ.
c) The charged exergy is 320 GJ.

In the normalized version of Table 9.1 shown in Table 9.2, the energy loss and exergy loss expressed as a percentage of the charged energy and exergy, respectively, are given as a function of the ratio of discharge time to time final temperature is reached, t_d/t_f. The discharge energy efficiency and discharge exergy efficiency are also shown against this time ratio in Table 9.2. Note that the recovered energy and exergy percentages correspond to the energy and exergy efficiencies, respectively.

The deviation between energy and exergy efficiencies is attributable to temperature differences between the charging and discharging fluids. As the discharging time increases, the deviation increases because the temperature of recovered heat decreases (Table 9.1 and Table 9.2). The energy efficiency reaches approximately 70% and the exergy efficiency 40% at the completion of discharging, even though the efficiencies are both 0% when discharging starts.

Note that if a threshold temperature is introduced (set to the actual temperature at the end of the discharge period of 5.24 days, i.e., 38 °C), the data in Table 9.1 and Table 9.2 for $t_d = 5.24$ days apply. Then, the discharged exergy (127 GJ) is 91% of the exergy recoverable in infinite time (139 GJ), but the discharged energy (1600 GJ) is only 83% of the maximum recoverable energy (1920 GJ). Furthermore, the exergy efficiency 40% is near the exergy efficiency attainable in infinite time (43.5%), although the energy efficiency (58%) remains greatly below the ultimate energy efficiency attainable (70%). Correspondingly, the exergy loss (190 GJ) exceeds the exergy loss in infinite time (180 GJ) by 5.5%, while the energy loss (1140 GJ) exceeds the energy loss in infinite time (820 GJ) by 39%.

The results suggest that aquifer TES performance measures based on exergy are more useful and meaningful than those based on energy, largely because exergy efficiencies account for the temperatures associated with energy transfers to and from the aquifer, as well as the quantities of energy transferred, while energy efficiencies account only for latter. Energy efficiencies can be misleadingly high, especially when heat is recovered at temperatures too low to be useful.

9.3   Analysis of a Ground-Source Heat Pump System

Energy and exergy analyses of a hybrid GSHP system, which uses a cooling tower for heat rejection, are presented in this section. The system considered is conceptual in nature, and the material presented here draws extensively on earlier reports (e.g., Lubis *et al.* 2011).

9.3.1   System Description and Operation

The hybrid GSHP system considered is shown in Figure 9.3. It has three main loops:

- A common interior building heat pump loop provides heating to the building and its environmental zones. That loop contains a water-source heat pump. The interior

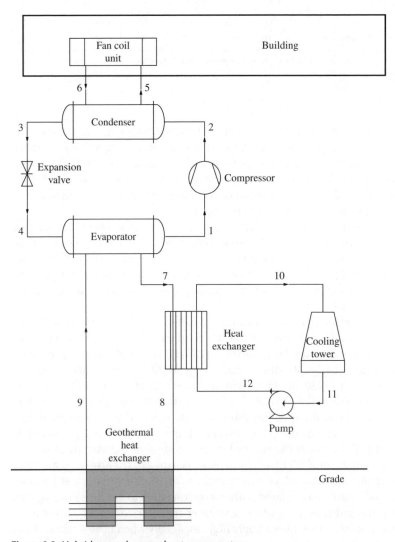

Figure 9.3 Hybrid ground-source heat pump system.

building heat pump loop serves as a heat source or sink, depending on the heating or cooling demands. The working fluid in the heat pump is R-134a.

- The interior building heat pump loop is connected to a ground heat exchanger loop. The ground provides a heat source and heat sink for the overall system.
- A cooling tower is connected in series with the ground heat exchanger in the hybrid system, in a cooling tower loop. The cooling tower serves as a supplemental heat rejection device and is used to ensure the energy flows to and from the ground are in balance over the year. The cooling tower loop is isolated from the building and ground piping loops with a plate heat exchanger.

Table 9.3 Process data for material streams in the hybrid ground-source heat pump system and the reference environment.

Stream[a]	Fluid description		Intensive thermodynamic properties					Extensive thermodynamic properties	
	Fluid	Phase	Temperature, T (°C)	Pressure, P (kPa)	Specific enthalpy, h (kJ/kg)	Specific entropy, s (kJ/kg K)	Specific flow exergy, ex (kJ/kg)	Mass flow rate, $\dot{m}$ (kg/s)	Exergy rate, $\dot{Ex}$ (kJ/s)
Refrigerant streams									
0	R-134a	Superheated vapor	2.2	101.3	257.3	1.039	0.000	—	—
1	R-134a	Superheated vapor	2.5	307	252.3	0.9346	23.57	0.02	0.47
2	R-134a	Superheated vapor	54.6	1011	287.5	0.9663	50.01	0.02	1.00
3	R-134a	Liquid	22.8	1011	83.3	0.3129	25.77	0.02	0.52
4	R-134a	Mixture	1.3	307	83.3	0.3189	24.11	0.02	0.48
Water streams									
0'	H_2O	Liquid	2.2	101.3	9.336	0.0337	0.000	—	—
5	H_2O	Liquid	75.0	250	314.2	1.016	34.48	0.021	0.71
6	H_2O	Liquid	27.8	250	116.7	0.406	4.836	0.021	0.10
10	H_2O	Liquid	8.0	220	33.84	0.1213	0.3718	1.07	0.40
11	H_2O	Liquid	11.0	200	46.41	0.1659	0.6768	1.07	0.73
12	H_2O	Liquid	11.0	240	46.47	0.1659	0.7175	1.07	0.77
Salinated water streams									
7	Brine	Liquid	13.0	250	54.84	0.1952	1.015	0.403	0.41
8	Brine	Liquid	5.0	250	21.27	0.07625	0.2082	0.403	0.08
9	Brine	Liquid	15.0	250	63.22	0.2244	1.359	0.403	0.55

a) Stream numbers correspond to Figure 9.3 except for "streams" 0 and 0' which represent the reference environment.

Table 9.4 Work and heat rates for the devices in the hybrid ground-source heat pump system.

Device	Heat transfer rate (kW)	Work input rate (kW)
Compressor	—	0.70
Condenser	4.1	—
Expansion valve	—	—
Evaporator	3.4	—
Plate heat exchanger	13.5	—
Pump	—	0.06

Note that dedicated water-to-water heat pumps may also be connected to the common interior building heat pump loop to meet domestic water heating needs for the building.

Values are presented in Table 9.3 for selected properties (mass flow rate, temperature, pressure) for the flows in the main loops (the R-134a heat pump working fluid, water, and brine) in the hybrid GSHP system. The state points in Table 9.3 correspond to the points in Figure 9.3.

The thermodynamic properties are either specified or obtained with software (Klein 2010).

The mechanical work and heat transfer rates associated with the components are given in Table 9.4.

The system is taken to operate in the following ambient conditions: a temperature of 2.2 °C and a pressure of 101.325 kPa.

9.3.2 Analyses

Energy and exergy analyses of the hybrid GSHP system are performed following the approach laid out by Dincer and Rosen (2007). Mass, energy, and exergy balances are applied to the system and its main parts to determine energy and exergy efficiencies, as well as the rates of exergy loss and exergy destruction.

The hybrid GSHP system is assessed while operating at steady-state. Kinetic and potential energies and work terms are neglected as is chemical exergy since it is not relevant to the analyses.

The reference environment for exergy analysis is taken to be at the ambient conditions specified above.

9.3.3 Analyses of Overall System

The coefficient of performance (COP) for the overall system is evaluated as the ratio of the heat transfer rate in the condenser (Figure 9.3) to the total energy input rate to the system in the form of work:

$$COP_{sys} = \frac{\dot{Q}_{cond}}{\dot{W}_{comp} + \dot{W}_{pump}} \tag{9.23}$$

The COP provides an energy-based measure of performance.

A corresponding exergy efficiency for the overall system can be written as the ratio of the thermal exergy transfer rate in the condenser to the total exergy input rate to the system:

$$\psi_{\mathrm{sys}} = \frac{\dot{Ex}_{\mathrm{out}}}{\dot{Ex}_{\mathrm{in}}} = \frac{\dot{Ex}_2 - \dot{Ex}_3}{\dot{W}_{\mathrm{comp}} + \dot{W}_{\mathrm{pump}}} \tag{9.24}$$

9.3.3.1 Analyses of Primary Mechanical Devices

The primary mechanical devices in the system are the compressor, the pump, and the expansion valve.

The compressor and the pump behave somewhat similarly and thus can be assessed in a common manner. That is, the exergy efficiency of these two devices can be evaluated as the ratio of the rate of exergy increase of the stream being increased in pressure to the rate of exergy input as work to the device.

The exergy efficiency can thus be written for the compressor as:

$$\psi_{\mathrm{comp}} = \frac{\dot{Ex}_2 - \dot{Ex}_1}{\dot{W}_{\mathrm{comp}}} \tag{9.25}$$

and for the pump as:

$$\psi_{\mathrm{pump}} = \frac{\dot{Ex}_{12} - \dot{Ex}_{11}}{\dot{W}_{\mathrm{pump}}} \tag{9.26}$$

The expansion valve is a device in which the pressure of the refrigerant is decreased with no product output (like mechanical power). The expansion valve is a dissipative device with no logical product. Hence it is not possible to evaluate a conventional efficiency for it. However, a somewhat meaningful exergy efficiency can nevertheless be written for the expansion valve as the ratio of its exergy output rate to exergy input rate:

$$\psi_{\mathrm{valve}} = \frac{\dot{Ex}_4}{\dot{Ex}_3} \tag{9.27}$$

The difference between the numerator and denominator here represents the exergy destruction rate in the expansion valve. Note that the exergy efficiency of the expansion valve can alternatively be expressed as follows:

$$\psi_{\mathrm{valve}} = \frac{\dot{Ex}_{\mathrm{out}}}{\dot{Ex}_{\mathrm{in}}} = 1 - \frac{\dot{Ex}_{\mathrm{dest}}}{\dot{Ex}_{\mathrm{in}}} = 1 - \frac{\dot{Ex}_3 - \dot{Ex}_4}{\dot{Ex}_3} \tag{9.28}$$

With this definition, the exergy efficiency of the expansion valve is zero, which can be viewed as more meaningful since all of the expended exergy is lost, reflecting the highly irreversible nature of the device.

9.3.3.2 Analyses of Primary Heat Exchange Devices

The primary heat exchange devices in the system are the heat exchanger, the condenser, and the evaporator. Like some of the mechanical devices, they behave similarly and thus can be assessed in a common manner.

All of the heat exchange devices have an input cold stream which is heated by a separate input hot stream. The exergy efficiency of each heat exchange device is evaluated as the ratio of the rate of exergy increase of the cold stream to the rate of exergy decrease of the hot stream.

With this approach, the exergy efficiency can be written for the condenser as:

$$\psi_{\text{cond}} = \frac{\dot{Ex}_5 - \dot{Ex}_6}{\dot{Ex}_2 - \dot{Ex}_3} \tag{9.29}$$

for the evaporator as:

$$\psi_{\text{evap}} = \frac{\dot{Ex}_9 - \dot{Ex}_7}{\dot{Ex}_4 - \dot{Ex}_1} \tag{9.30}$$

and for the plate heat exchanger as:

$$\psi_{\text{PHX}} = \frac{\dot{Ex}_7 - \dot{Ex}_8}{\dot{Ex}_{12} - \dot{Ex}_{10}} \tag{9.31}$$

9.3.4 Performance

9.3.4.1 Exergy Destruction Rates
Exergy destruction rates for the hybrid GSHP are listed in Table 9.5.

The greatest exergy destruction rates occur in the compressor and the condenser. The exergy destruction rates in these components are caused by friction and heat transfer across a finite and relatively large temperature difference.

The exergy efficiency of the expansion valve evaluated based on Equation (9.27) is high (93.6%), reflecting the fact that most of the exergy entering the expansion valve exits.

9.3.4.2 Efficiencies
Exergy efficiencies for the hybrid GSHP are listed in Table 9.5. The COP of the system is found to be 5.34 and the exergy efficiency 63.4%. These values are high but not unreasonable.

It is instructive to compare the COP and exergy efficiency for the present system with corresponding values for other similar systems. In one separate assessment, GSHP systems have been shown to have an exergy efficiency of 68.1% and a COP of 2.27–3.14

Table 9.5 Exergy destruction rates and efficiencies for the overall hybrid ground-source heat pump system and its devices.

Device	Exergy destruction rate		Exergy efficiency (%)
	Absolute (kW)	Relative (% of total)	
Compressor	0.17	33	75
Condenser	0.13	24	79
Expansion valve	0.03	6	94
Evaporator	0.13	24	8
Plate heat exchanger	0.05	9	88
Pump	0.02	3	71
Overall system	0.53	100	63

(Ozgener and Hepbasli 2007a). In another assessment, a GSHP systems was shown to have a COP of 3.64 (Ozgener and Hepbasli 2007b). The higher COP for the latter case is mainly associated with the relatively higher temperature of geothermal water with respect to the ground.

The comparative data suggest that hybrid GSHPs can achieve higher COPs than ground- and air-source heat pumps, as the COP values reported such systems do not exceed four (Akpinar and Hepbasli 2007; Kuzgunkaya and Hepbasli 2007a, b; Ozgener and Hepbasli 2007a, b; Bi *et al.* 2009). Incorporating another source of thermal energy [e.g., solar energy (Kuzgunkaya and Hepbasli 2007b) or geothermal water] into a system may allow for higher COPs.

The results of the energy and exergy analyses of a hybrid GSHPg system suggest that hybrid GSHP systems operate at higher COP and exergy efficiencies than typical air-source heat pump systems. The latter typically exhibit COPs of 1–3 and exergy efficiencies less than 30% (Sanner 2008).

9.4   Analysis of a System Integrating Ground-Source Heat Pumps and Underground Thermal Storage

Energy and exergy analyses are described of a system integrating GSHPs and underground thermal storage. The application considered is the north campus of the University of Ontario Institute of Technology, located in Oshawa, Ontario, Canada. Most of the university's buildings in its north campus were designed to be heated and cooled using GSHPs in concert with underground thermal energy storage. This system has already been introduced in this book (see Section 4.4.5) and is described in greater detail subsequently as a case study (see Section 12.5). This analysis draws heavily on that presented by Dincer and Rosen (2011, 2013) and Kizilkan and Dincer (2012).

9.4.1   Rationale for Using a System Integrating Ground-Source Heat Pumps and Thermal Storage

The rationale for using GSHPs and underground thermal storage at the university was to achieve reduced energy resource use, environmental emissions and lifetime financial costs, compared with more conventional heating and cooling systems. GSHPs can also facilitate demand-side management for heating and cooling applications in residential and commercial buildings.

9.4.2   Description of a System Integrating Ground-Source Heat Pumps and Thermal Storage

The integrated system with GSHPs and thermal storage is illustrated in Figure 9.4.

The main components of the integrated system, along with the terms used to denote them throughout the analysis, are as follows:

- borehole heat exchanger (BHE)
- evaporator (evap)
- fan coil (FanCoil)
- condenser (cond)

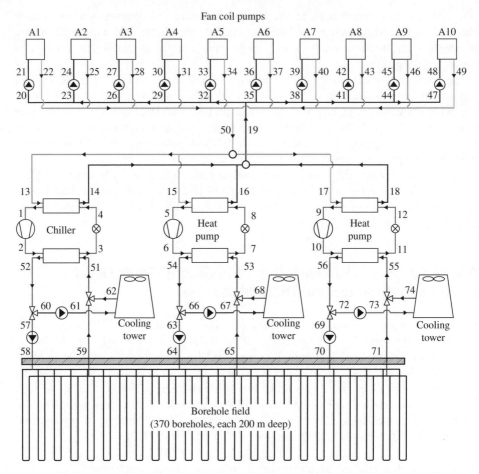

Figure 9.4 Schematic of ground-source heat pump and borehole thermal energy storage system. Source: Adapted from Dincer and Rosen (2013) and Kizilkan and Dincer (2012).

- compressor (comp)
- expansion valve (ExpV)
- circulating pump (Pump)

The first four are the primary heat transfer devices, and the last three are the main mechanical components.

9.4.2.1 Integrated System

The university's central plant houses and operates the cooling and heating system for the campus (including the underground thermal storage). Variable displacement centrifugal chillers with magnetic bearings, which provide very good part-load performance, are used to provide energy. Pumps convey the working fluid between the buildings and the underground thermal storage. Heat pump modules assist in cooling. Chilled water is supplied from two multi-stack chillers, each having seven modules, and two sets of heat pumps each with seven modules. The condenser water enters the underground thermal storage, which retains the heat from the condensers for use in the winter (when the heat pumps reverse) and provides low-temperature hot water for the campus (Dincer and Rosen 2011).

9.4.2.2 Underground Storage

The type of underground thermal storage used at the University of Ontario Institute of Technology is borehole thermal energy storage (BTES). This university's geothermal well field is the largest and deepest in Canada and one of the largest in North America (Dincer and Rosen, 2011). The BTES system uses the ground as a heat source or sink or storage medium, and meets the heating and cooling load of most of the buildings in the university's north campus in concert with GSHPs.

The BTES contains many borehole heat exchangers (BHEs), which are boreholes in the ground that contain a tube through which a heat transfer fluid flows. Although the space between a borehole wall and the ground can be filled with groundwater or grout, the Swedish practice of water-filled BHEs was utilized instead of the North American practice of grouted BHEs.

Test drilling programs were carried out to determine the feasibility of thermal storage in the overburden and bedrock formations at the site, and to aid design. The thermal conductivity test results suggested that a field of 370 boreholes, each 200 m in depth, would meet the energy demand.

9.4.3 Cooling Mode Operating Data of a System Integrating Ground-Source Heat Pumps and Thermal Storage

The university was originally designed for 10 buildings, and the analyses in this section consider the original design in the cooling mode. Note that the university in actuality now has a greater number of buildings, split among its two campuses. Actual cooling load data for the system are utilized in the assessment. The total cooling load is about 7000 kW for the university's north campus buildings. The system illustration in Figure 9.4 shows it as it operates in cooling mode.

A glycol solution, encased in polyethylene tubing, circulates through an interconnected, underground network. A 15% glycol solution is the source that is circulated through the BHE mounted in the ground. Inlet and outlet temperatures of solution to and from the ground are 29.4 and 35 °C, respectively. The glycol solution concentration is 30% and is circulated between the system and buildings to transfer heat. Inlet and outlet temperatures of the solution to and from the fan coils are 5.5 and 14.4 °C, respectively. During the winter, fluid circulating through tubing extended into the wells collects heat from the earth and carries it into the buildings. In summer, the system reverses to extract heat from the buildings and transmits it to the ground.

Operating data, in terms of temperature, mass flow rate and exergy for streams in the system (Figure 9.4) are listed elsewhere (Kizilkan and Dincer 2012; Dincer and Rosen 2013).

9.4.4 Analysis of a System Integrating Ground-Source Heat Pumps and Thermal Storage

The analysis of the system integrating GSHPs and thermal storage is carried out for steady-state and steady flow conditions.

The reference-environment temperature and pressure are fixed at 24 °C and 98.825 kPa, respectively.

In terms of notation conventions, heat transfer to the system and work transfer from the system are taken to be positive throughout the analysis.

9.4.4.1 Assumptions and Simplifications

Various assumptions and simplifications are invoked to facilitate the analysis and permit illustrative results without requiring excessive detail.

For the overall analysis:

- Potential and kinetic energy effects are negligible.
- It is assumed that there are no chemical or nuclear reactions.
- The reference-environment temperature and pressure are assumed to be constant, even though they vary throughout the year.

Assumptions and simplifications are also made for system components:

- The isentropic efficiencies are assumed to be 85% for the chiller and 80% for the heat pumps.
- The compressor is assumed to be adiabatic.
- The mechanical efficiency is assumed to be 80% and the electrical efficiency 84% for the compressor.
- The mechanical efficiency is assumed to be 85% and the electrical efficiency 88% for the circulating pumps.
- Electrical power inputs to the fan-coil fans are assumed to be negligible.
- The chiller system is taken to use refrigerant R507A as its working fluid.
- The heat pump is taken to use refrigerant R407C as its working fluid.

9.4.4.2 Exergy Destruction Rates and Relative Irreversibilities

Exergy destruction (or irreversibility) rates for the overall integrated system in Figure 9.4 and its primary components can be determined by writing exergy balances.

Exergy destruction rates are first written for the primary heat transfer components in the overall integrated system. For the BHE, which interfaces with the underground thermal energy storage:

$$\dot{Ex}_{\text{dest,BHE}} = (\dot{Ex}_{\text{BW,in}} - \dot{Ex}_{\text{BW,out}}) - \dot{Q}_{\text{BHE}}\left(1 - \frac{T_0}{T_{\text{BHE}}}\right) \tag{9.32}$$

For the evaporator (evap) and condenser (cond), respectively:

$$\dot{Ex}_{\text{dest,evap}} = (\dot{Ex}_{\text{evap,in}} - \dot{Ex}_{\text{evap,out}}) + (\dot{Ex}_{\text{CW,in}} - \dot{Ex}_{\text{CW,out}}) \tag{9.33}$$

and

$$\dot{Ex}_{\text{dest,cond}} = (\dot{Ex}_{\text{cond,in}} - \dot{Ex}_{\text{cond,out}}) + (\dot{Ex}_{\text{BW,in}} - \dot{Ex}_{\text{BW,out}}) \tag{9.34}$$

For the fan coil (FanCoil):

$$\dot{Ex}_{\text{dest,FanCoil}} = (\dot{Ex}_{\text{CW,in}} - \dot{Ex}_{\text{CW,out}}) + \dot{Q}_{\text{FanCoil}}\left(1 - \frac{T_0}{T_{\text{FanCoil}}}\right) \tag{9.35}$$

Exergy destruction rates are first written for the primary mechanical components in the overall integrated system are now considered. For the compressor (comp) and circulating pump (pump), respectively, both of which serve to increase the pressure of flows:

$$\dot{Ex}_{dest,comp} = \dot{Ex}_{comp,in} - \dot{Ex}_{comp,out} + \dot{W}_{comp} \tag{9.36}$$

and

$$\dot{Ex}_{dest,pump} = \dot{Ex}_{pump,in} - \dot{Ex}_{pump,out} + \dot{W}_{pump} \tag{9.37}$$

For the expansion valve (ExpV), which is a dissipative device (i.e., it provides no useful product or output):

$$\dot{Ex}_{dest,ExpV} = \dot{Ex}_{ExpV,in} - \dot{Ex}_{ExpV,out} \tag{9.38}$$

The relative irreversibility can be evaluated for each component as the ratio of its exergy destruction rate to the overall exergy destruction rate for the integrated system. The relative irreversibilities provide normalized contributions of each component to the overall system's irreversibility.

Exergy destruction rates and relative irreversibilities are listed in Table 9.6 for the overall system and each of its main component groups. In that table, the division between heat transfer and mechanical devices is also shown. The overall exergy destruction rate is found to be 1346 kW. The greatest exergy destruction rates in the integrated system occur in the compressors of the chiller and heat pumps, the condensers, the expansion valves, and the evaporators.

Table 9.6 Exergy destruction rates and relative irreversibilities for the overall integrated system and its main component groups.

Component group	Exergy destruction rate (kW)	Relative irreversibility (%)
Heat exchange devices		
Borehole heat exchangers (1–3)	68	5
Condensers (1–3)	333	25
Evaporators (1–3)	290	22
Fan coils (A1–A10)	68	5
Cooling towers (1–3)	70	5
Subtotal	**829**	**62**
Mechanical devices		
Compressors (1–3)	192	14
Fan-coil pumps (A1–A10)	10	0.7
Borehole heat exchanger pumps (1–3)	<0.1	<0.01
Cooling tower pumps (1–3)	5	0.4
Expansion valves (1–3)	311	23
Subtotal	**518**	**38**
Total	**1346**	**100**

9.4.4.3 Overall Efficiencies

Exergy efficiencies are determined for the overall system and its components. The exergy destruction rate balances from the previous subsection can be employed to assist in writing the efficiencies.

For the overall system, the exergy efficiency can be written as:

$$\psi_{sys} = \frac{\dot{Ex}_{out}}{\dot{Ex}_{in}} = \frac{\sum \dot{Ex}^{\dot{Q}}_{FanCoil}}{\sum \dot{W}_{comp} + \sum \dot{W}_{pump}} \tag{9.39}$$

9.4.4.4 Component Efficiencies

Exergy efficiencies are first written for the primary heat transfer components in the overall integrated system. For the BHE, the evaporator (evap), the condenser (cond), and the fan coil (FanCoil), respectively:

$$\psi_{BHE} = \frac{\dot{Ex}_{BW,in} - \dot{Ex}_{BW,out}}{\dot{Ex}^{\dot{Q}}_{BHE}} \tag{9.40}$$

$$\psi_{evap} = \frac{\dot{Ex}_{CW,in} - \dot{Ex}_{CW,out}}{\dot{Ex}_{cond,out} - \dot{Ex}_{cond,in}} \tag{9.41}$$

$$\psi_{cond} = \frac{\dot{Ex}_{BW,out} - \dot{Ex}_{BW,in}}{\dot{Ex}_{cond,in} - \dot{Ex}_{cond,out}} \tag{9.42}$$

$$\psi_{FanCoil} = \frac{\dot{Ex}_{CW,out} - \dot{Ex}_{CW,in}}{\dot{Ex}^{\dot{Q}}_{FanCoil}} \tag{9.43}$$

Exergy destruction rates are next written for the primary mechanical components in the overall integrated system. For the compressor (comp), the circulating pump (pump), and the expansion valve (ExpV), respectively:

$$\psi_{comp} = \frac{\dot{Ex}_{comp,out} - \dot{Ex}_{comp,in}}{\dot{W}_{comp}} \tag{9.44}$$

$$\psi_{pump} = \frac{\dot{Ex}_{pump,out} - \dot{Ex}_{pump,in}}{\dot{W}_{pump}} \tag{9.45}$$

$$\psi_{ExpV} = \frac{\dot{Ex}_{ExpV,out}}{\dot{Ex}_{ExpV,in}} \tag{9.46}$$

Although the latter component is dissipative, an exergy efficiency can be written, but it reflects simply the ratio of output exergy rate to input exergy rate for the device.

Exergy efficiencies are listed in Table 9.7 for the overall system and each of its component groups. The exergy efficiency for the BTES system on a product/fuel basis is found to be 62%.

The efficiencies described in this case study are consistent with those of Hepbasli (2005), who found, for a GSHP-based district heating system, with U-tube ground heat exchangers, the heating coefficient of performance of the heat pump to be 2.85 and the exergy efficiencies for the heat pump and overall system, respectively, to be 66.8 and 66.6%.

Table 9.7 Exergy efficiencies for the overall integrated system and its main component groups.

Component group	Mean exergy efficiency (%)
Heat exchange devices	
Borehole heat exchangers (1–3)	42
Condensers (1–3)	28
Evaporators (1–3)	45
Fan coils (A1–A10)	69
Cooling towers (1–3)	55
Mechanical devices	
Compressors (1–3)	87
Fan-coil pumps (A1–A10)	7
Borehole heat exchanger pumps (1–3)	3
Cooling tower pumps (1–3)	3
Expansion valves (1–3)	86
Overall system	62

Table 9.8 Variations in exergy destruction rates and efficiencies for the overall integrated system with selected design parameters.

Design parameter	Exergy destruction rate (kW)	Exergy efficiency (%)
Evaporator temperature (°C)		
5	1010	66.9
3	1100	66.0
1	1150	65.5
−1	1230	64.7
−3	1290	64.0
−5	1390	63.4
Condenser temperature (°C)		
70	2790	58.9
65	2140	60.0
60	1840	61.2
55	1520	62.4
50	1310	63.5
45	1200	65.0
Inlet glycol solution temperature (°C)		
34	1950	56.9
32	1340	63.9
30	1340	63.8
28	1350	63.5

9.4.4.5 Variations of Exergy Destruction Rates and Efficiencies with Key Design Parameters

The variations of the exergy destruction rates and efficiencies with several key design parameters in the system are considered here.

The variations of overall system exergy destruction rate and efficiency with evaporator temperature are shown in Table 9.8 (top portion). The overall exergy destruction rate decreases and the exergy efficiency increases, both in a near-linear manner, with increasing of evaporator temperature.

The variations of overall system exergy destruction rate and efficiency with condenser temperature are shown in Table 9.8 (middle portion). The overall exergy destruction rate decreases and the overall exergy efficiency increases in a near-linear manner with decreasing condenser temperature.

The variations of the exergy destruction rate and efficiency with the glycol solution temperature entering the refrigeration system are shown in Table 9.8 (bottom portion). The glycol solution temperature entering the refrigeration system (i.e., the condenser) is higher than the ground temperature in summer, because of heat rejection from the circulating glycol solution to the ground. It is seen that the system exergy destruction rate increases as the entering glycol solution temperature decreases.

References

Akpinar, E.K. and Hepbasli, A. (2007) Comparative study on exergetic assessment of two ground-source (geothermal) heat pump systems for residential applications. *Building and Environment*, **42**, 2004–2013.

Bi, Y., Wang, X., Liu, Y. *et al.* (2009) Comprehensive exergy analysis of a ground-source heat pump system for both building heating and cooling modes. *Applied Energy*, **86**, 2560–2565.

Dincer, I. and Rosen, M.A. (2007) *Exergy: Energy, Environment, and Sustainable Development*, Elsevier, Oxford.

Dincer, I. and Rosen, M.A. (2011) *Thermal Energy Storage: Systems and Applications*, 2nd edn, John Wiley & Sons, Ltd, Chichester.

Dincer, I. and Rosen, M.A. (2013) *Exergy: Energy, Environment, and Sustainable Development*, 2nd edn, Elsevier, Oxford.

Hepbasli, A. (2005) Thermodynamic analysis of a ground-source heat pump system for district heating. *International Journal of Energy Research*, **29**, 671–687.

Hoyer, MC, Walton, M, Kanivetsky, R & Holm, TR 1985 Short-term aquifer thermal energy storage (ATES) test cycles, St. Paul, Minnesota, U.S.A. Proceedings of the 3rd International Conference on Energy Storage for Building Heating and Cooling, September 22–26, 1985, Toronto, Canada. Public Works Canada, pp. 75–79.

Kizilkan, O. and Dincer, I. (2012) Exergy analysis of borehole thermal energy storage system for building cooling applications. *Energy and Buildings*, **49**, 568–574.

Klein SA 2010 *Engineering Equation Solver (EES)*. Academic Commercial, F-Chart Software. www.fChart.com (accessed July 27, 2016).

Kuzgunkaya, E.H. and Hepbasli, A. (2007a) Exergetic evaluation of drying of laurel leaves in a vertical ground-source heat pump drying cabinet. *International Journal of Energy Research*, **31**, 245–258.

Kuzgunkaya, E.H. and Hepbasli, A. (2007b) Exergetic performance assessment of a ground-source heat pump drying system. *International Journal of Energy Research*, **31**, 760–777.

Lubis, L.L., Kanoglu, M., Dincer, I. and Rosen, M.A. (2011) Thermodynamic analysis of a hybrid geothermal heat pump system. *Geothermics*, **40**, 233–238.

Ozgener, O. and Hepbasli, A. (2007a) A parametrical study on the energetic and exergetic assessment of a solar assisted vertical ground-source heat pump system used for heating a greenhouse. *Building and Environment*, **42**, 11–24.

Ozgener, O. and Hepbasli, A. (2007b) Modeling and performance evaluation of ground source (geothermal) heat pump systems. *Energy and Buildings*, **39**, 66–75.

Rosen, M.A. (1999) Second-law analysis of aquifer thermal energy storage systems. *Energy – The International Journal*, **24**, 167–182.

Sanner, B 2008 Ground source heat pumps: Development and status. Ground Coupled Heat Pumps of High Technology Final Workshop, May 5, 2008, Berlin, Germany, presentation 09–10.

10

Environmental Factors

10.1 Introduction

In addition to superior thermal performance and cost effectiveness as factors analyzed when making decisions on using energy systems, environmental aspects are being increasingly considered. For instance, regulations exist or are being established in many jurisdictions to reduce CO_2 emissions and protect flora and fauna. From an environmental perspective, preserving natural habitats in the ground by avoiding drastic changes in temperature and moisture content are important factors, especially when considering the installation of ground heat exchangers (GHEs) at various depths in the ground. While the use of geothermal systems is widespread, having had a revival in the 1980s and recently, the impact of these systems on the environment is now being questioned. Due to their efficiency, the use of geothermal energy should be encouraged. However, little research is available to guide regulatory agencies and industry towards designs and installations that maximize their sustainability and minimize possible environmental impacts. On the positive side, prevention of greenhouse gas (GHG) emissions by the use of such clean energy systems should also be carefully examined. In this chapter, environmental benefits and impacts of geothermal systems are studied.

10.2 Environmental Benefits

One of the main environmental advantages of the use of ground-source heat pump (GSHP) systems is their low wastes/emissions. In some cases, they can achieve emission-free operation. Furthermore, compared with petroleum-based energy systems, GSHP systems do not have the same risks in such areas as transportation, storage, operation, and groundwater contamination.

The waste emissions of GSHPg systems can be evaluated. Determinations of reductions in CO_2 and other emissions achieved by GSHP systems normally involve comparisons with the emissions of conventional heating and/or cooling systems, based on the hypothesis that the GSHP system replaces the conventional system. Conventional systems range from heaters that use fossil fuels directly, such as gas burners, to those that use fossil fuels indirectly, such as heaters that use electricity supplied from fossil fuel burning power plants. Coupling of GSHPs to an electricity grid to heat and cool buildings usually results in a reduction of CO_2 and other emissions compared with conventional space heating technologies that use fossil fuels.

Geothermal Energy: Sustainable Heating and Cooling Using the Ground, First Edition.
Marc A. Rosen and Seama Koohi-Fayegh.
© 2017 John Wiley & Sons, Ltd. Published 2017 by John Wiley & Sons, Ltd.

Since emissions associated with electricity generation can be reduced by using more efficient generation technologies, such as cogeneration plants, the emissions associated with the use of GSHPs also decrease through the use of high-efficiency electricity generation. These emissions associated with GSHPs can be eliminated when electricity supplied to the heat pump is generated from renewable energy resources such as solar, wind and high-temperature geothermal materials.

In countries that generate electricity from nuclear and renewable energy sources, as well as fossil fuels, the partial amounts of CO_2 emissions from each source must be carefully examined to determine the mass of emitted CO_2 and other emissions per unit of electricity generated. The same reasoning applies for the evaluation of emissions for GSHPs in a country that generates electricity from various sources.

In addition, an emissions analysis is dependent on factors such as the seasonal coefficient of performance (COP) of the heat pump, especially if it is to replace conventional heating and cooling systems. The number of hours that a GSHP system operates once installed to cover the heating and cooling loads of communities is an important factor in determining the emissions. Such information helps predict how much electricity the GSHP uses per year to heat and cool buildings regardless of its electricity source. In order to accurately predict such factors, experimental or simulated heating and cooling load data from actual buildings, climate and weather data, and average ground temperatures are needed. Since such analyses are relatively complex, many studies, especially those on emissions for a large number of buildings, often use assumptions for the average COP of GSHPs and the number of hours they are in full load operation every day. In other cases, more complete simulations and analyses of single buildings, including simulations of their heating and cooling loads using regional weather data, are used to predict emission reductions from using GSHPs.

Much effort has focused on evaluating the reductions in CO_2 emissions for GSHPs, given the contribution such emissions make to climate change. In the remainder of this section, data from several studies of CO_2 emissions reduction reported in the literature are discussed.

In a recent study, 10 Indian states with populations ranging from 0.6 to 12.5 million are considered in an analysis to determine the reduction in CO_2 emissions using GSHPs for heating and cooling (Sivasakthivel *et al*. 2014). It is assumed that heating and cooling is otherwise provided using conventional electric heaters and air conditioners. The results indicate that the power demand for electric space heating ranges from 1416 to 7085 GW over the year. This demand reduces by 67–80% to about 471–1416 GW if GSHPs are used. Moreover, the electrical power use by a conventional air conditioner for cooling purposes is about 5506–27 532 GW for the 10 states. Reductions in power demand of 13–48% to 4811–14 440 GW are expected if GSHPs are used for cooling. Considering heating and cooling energy needs, the CO_2 emissions decline from 5270 to 26 352 million kg of CO_2 associated with conventional systems for space heating and cooling to 4022–12 071 million kg using GSHP, a reduction of about 24–54% per year. Such reductions in CO_2 emissions become more pronounced if the electricity for the GSHP operation is generated from renewable energies.

In Germany, renewable and nuclear sources are also used for electricity generation along with the fossil fuels. It is thus expected that emissions associated with the use of GSHPs are even lower than for countries deriving 100% of their electricity from fossil fuels. For an average German electricity mix of 12% renewable energy, 61% fossil fuels, and 27% nuclear energy, a CO_2 equivalent of 594 g CO_2/kWh was used in a study (Blum *et al.* 2010) to calculate the CO_2 emissions associated with using a GSHP for heating. The GSHP that replaces the conventional heating and cooling systems is taken to have a COP of 4 and to operate 2000 h per year. We consider the CO_2 emissions of conventional heating systems as the base case against which GSHP systems are compared. A CO_2 equivalent for an energy heating mix for residential buildings of 53% natural gas, 42% heating oil, 4% electricity, and 1% coal is chosen. The mass of CO_2 emitted per unit electricity use for heating for this scenario is found to be 149 g CO_2/kWh for GSHPs using the German electricity mix, compared with 229 g CO_2/kWh for the conventional heating mix, a reduction of 35% in CO_2 emissions.

Leong *et al.* (2011) simulate an existing 227 m^2 (2440 ft^2) old house located in Wasaga Beach, Ontario, Canada. The house is equipped with electric resistance heating. The annual energy consumption for electric heating is compared with simulation results of installing a GSHP using a horizontal ground heat exchanger (HGHE) in an optimum configuration that results in the best performance. These two scenarios are compared with another scenario that assumes the house is being heated via a natural gas furnace with an efficiency factor of 0.8 (including the effects of rated full-load efficiency, part-load performance, and over-sizing). The annual use of natural gas is found to be 3700 m^3 using a natural gas furnace with an efficiency factor of 0.8 (including the effects of rated full-load efficiency, part-load performance, and over-sizing) and 3510 m^3 for a GSHP that consumes electricity generated by natural gas. The comparison demonstrates the environmental benefits of operating the GSHP system, that is, an annual reduction in GHG emissions of 5% relative to operating a natural gas furnace for heating. The environmental benefits are greater if the electricity for the GSHP system is generated by green technologies, such as deep geothermal energy, solar and wind energy sources.

The number of GSHP systems is continuously increasing worldwide, helping to mitigate CO_2 emissions and their contributions to global climate change. The contributions of GSHP systems to reducing CO_2 emissions can be further enhanced through the integration of renewable energies such as solar, wind, and deep geothermal into electricity grids so they constitute a larger portion of the electricity generation mix.

10.3 Environmental Impacts

As with most human activities, the use of geothermal heat exchangers introduces the potential for causing environmental impacts. Such impacts have often been associated with CO_2 emissions when discussing energy systems. As mentioned in the previous section, the CO_2 emissions and impact of GSHP systems on the environment are directly dependent on the source of their electricity supply. Several other factors related to GSHPs are considered potential contributors to environmental impacts, such as heat pump manufacturing, leakage of the heat carrier fluids to ground, groundwater and aquifers, and thermal and chemical contamination of natural water resources and temperature-sensitive underground ecosystems. Some of these effects have been demonstrated to be minor for heat pump systems. Leakages of commonly used refrigerants in heat pumps represent GHG emissions and global warming potentials that are much higher than those of CO_2, on a per molecule basis.

Little research has been done regarding the impact of geothermal systems on the local environment, probably because they have been presumed to be on the whole environmentally beneficial. However, some environmental concerns have been noted and examined.

The movement of heat in the form of thermal plumes through the ground exhibits the potential for environmental impact and has been investigated. The migration of thermal plumes away from GSHP systems and changes in temperature induced by either closed- or open-loop systems, or due to changes in groundwater flow patterns from open-loop systems, may cause undesirable temperature rises in nearby temperature-sensitive ecosystems where small temperature differences are important. For example, temperature disturbances in the ground caused by the operation of GSHP systems may result in disruption to sensitive life stages of aquatic organisms and reduce reproduction rates. Similar environmental effects are observed for heat loops and waterline projects (i.e., those in rivers and lakes) (Fisheries and Oceans Canada 2009).

Markle and Schincariol (2007) investigate the potential thermal impacts from below-water-table aggregate extraction on a cool-water stream in Southwestern Ontario, Canada which supports Brook trout and cool-water micro-invertebrates. They demonstrate both the persistence of thermal plumes (persisting in an aquifer for 11 months and migrating up to 250 m along the gradient), and the sensitivity of the aquatic environment to very small temperature perturbations. Their results show that there is a narrow range for spawning in cold-water streams. These need to be cooled in the summer and warmed in the winter by groundwater flow. Once the groundwater temperature is affected due to the performance of GHEs, it can affect the temperature of the cold-water streams, making these sites unsuitable for spawning. A study on the effects of thermal fluctuations on microorganisms in aquifers having a geothermal well field show an increase in total microbial count in aquifer samples, which correlate with the increase in temperature in the geothermal well field (York *et al.* 1998). Moreover, counts of cultured bacteria suggest that, even when no significant differences in total bacterial number are observed, changes may occur in the types of microorganisms present in the aquifers of the geothermal well field.

In heating- or cooling-dominated climates, an annual energy imbalance is placed on the ground loop due to heating, cooling and hot water production. For example, Manitoba has a heating-dominated climate and there are concerns regarding the long-term thermal performance of ground loop heat pumps. Long-term thermal performance of such ground loop systems with unbalanced energy inputs and outputs in the ground may result in large temperature rises in the region where the loop is installed, as well as the migration of thermal plumes away from these systems and any associated environmental impacts.

Thermal imbalances can cause significant problems with a heat pump's long-term performance and sustainability if not properly considered at the design phase (Andrushuk and Merkel 2009).

It is not conclusively known at this point if the net environmental benefits of geothermal energy systems, when environmental impacts are also taken into account, are acceptable. That is, the environmental impacts of geothermal energy systems may be acceptable considering the fact that the technologies can reduce fossil fuel consumption and, therefore, lower GHG emissions. Of course, answers to these questions are also dependent on whether geothermal systems can be developed in the future in a manner that raises their efficiencies and reduces their potential for harming the environment.

Using models that are capable of simulating the heat exchange processes within a GSHP system and the surrounding environment and of performing local scale assessments, the migration of thermal plumes into the hydrogeological environment can be simulated and used to perform intermediate and regional scale assessment of environmental and ecological impact associated with GSHPs. That is the focus of the assessments in the next two subsections.

10.3.1 Environmental Assessment of a Horizontal Ground Heat Exchanger

In the example model presented in Section 8.4 for modeling various configurations of a horizontal ground heat exchanger (HGHE) for a single house, simulations by Leong *et al.* (2010) are performed for a single-layer HGHE (at 1.5 m below the ground surface) and a double-layer HGHE (at 1.0 m and 1.5 m below the ground surface) with a horizontal pipe spacing of $H = 0.5$ m (see Figure 8.15) operating over a 10-year period. These simulations can be utilized to verify the ground thermal and moisture sustainability.

To facilitate comparisons of the ground temperature and moisture content, the day before the heating season commences (i.e., September 19) is selected as a reference date. For natural background conditions (i.e., no GSHP system), the average ground temperature and moisture content between grade and 6.0 m below ground are 13.9 °C and 0.168 m^3 of water/m^3 of soil, respectively. If the ground temperature rises, the soil moisture content may be impacted, possibly harming underground biological life.

The results of the comparison for single- and double-layer HGHEs are considered separately:

- **Single-layer GHE.** On the reference day after 10 years of operation, the average ground temperature and moisture content for this case are 10.8 °C and 0.200 m^3 of water/m^3 of soil, respectively. Those new conditions are attained much earlier than 10 years, however, being observed roughly after 3 years of GSHP operation. Thus, the ground experiences, relative to its natural conditions, a temperature decrease of 3.1 °C and a small increase in moisture content for the single-layer GHE case. It is not anticipated that the increase in soil moisture content will be problematic. The ground temperature decrease relative to the background may or may not be problematic, although it is noted that the average ground temperature of 10.8 °C exceeds by a good margin the freezing temperature of water so icing is probably not an issue for the case of a single-layer GHE.
- **Double-layer GHE.** For this case, the average ground temperature and moisture content on the reference day after 10 years of operation are 13.1 °C and 0.200 m^3 of water/m^3 of soil, respectively. In this case, the temperature decline is only 0.8 °C while the ground moisture content again rises by a small amount so it remains similar to the natural conditions. It is not anticipated that the increase in soil moisture content and decrease in ground temperature will be problematic for the case of a double-layer GHE.

10.3.2 Environmental Assessment of a Borefield

A domain consisting of several borehole systems, each consisting of 16 vertical boreholes, is considered (see Figure 8.13). The borehole systems are placed at every 100 m and the boreholes are installed at 6-m distances. The details of calculation of the ground heat load profile are included in Section 6.3.

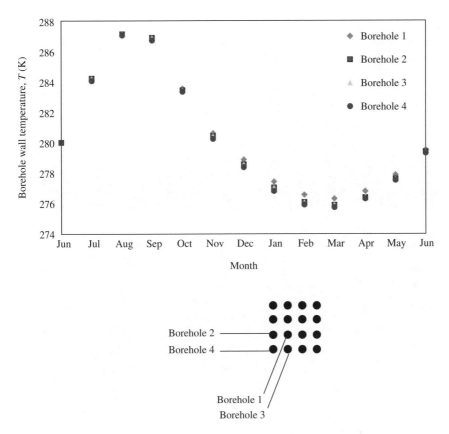

Figure 10.1 Annual temperature variations of the borehole wall.

The annual temperature variations of the borehole wall for the four boreholes are shown in Figure 10.1. It is seen that the temperature response of the borehole wall for the different borehole placements does not vary greatly (i.e., variations are less than 0.2%). This might be due to the stronger dependence of the borehole temperature on the heat flux on the wall than borehole placement. The slight variation that is noticed between the temperature rise of the boreholes is due to their relative location. The boreholes that are surrounded by other boreholes (Borehole 1) experience a slightly higher temperature in the heat delivery mode and, therefore, their temperature drop in the heat removal mode is lower than other boreholes. Comparing the temperature variations with the periodic heat flux variations at the borehole walls is illustrated in Figure 6.11. It is noticed that the temperature variation of the borehole wall exhibits some similarities to the transient variation of the heat flux, with its maximum and minimum values occurring in the second and ninth months, respectively.

Figure 10.2 shows the temperature contours in the ground surrounding the boreholes for 1 year. It is seen that the maximum temperature occurs in the ground immediately outside of the borehole wall and when the heat flux is at its maximum. Furthermore, it is noticed for the current problem that the thermal plume from the system reaches its furthest extent at the end of Month 10.

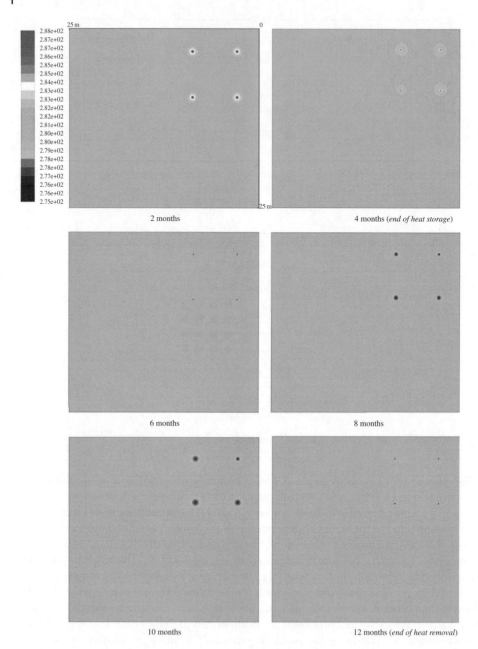

Figure 10.2 Temperature contours in the ground surrounding 4 of the 16 boreholes in the system (the holes surrounded with the highest temperature gradient are shown in the figure) in Year 1. (Note that four boreholes are shown here due to symmetry.)

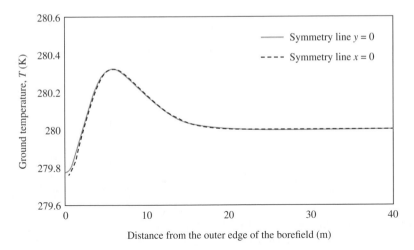

Figure 10.3 Ground temperature outside the borehole field after 10 months of system operation.

Figure 10.3 shows the ground temperature outside the borefield along the two lines of symmetry that intersect in the center of the borefield (see Figure 8.13a): from the borefield center, one line of symmetry is towards another borefield (symmetry $y=0$) and one is directed towards the farfield (symmetry $x=0$). It is seen that considering temperature rises of no more than 0.2 °C in the ground, the thermal plume in the ground caused by the system extends about 10 m from the outer edge of the borefield ($x=9$ m) and does not have any interaction with its neighboring system. It can be observed that, if the borehole spacing and the distance to neighboring borehole systems are kept to a certain distance, for a certain amount of heat flux, here given in Figure 6.11:

- There should not be any notable thermal interaction between neighboring systems.
- There should not be any thermal plume flowing away from the system.

It is also observed in Figure 10.3 that the temperature of the ground is reduced by about 0.2 °C at the borefield boundary, followed by a temperature rise of 0.3 °C in the 10-m region outside of the borefield. The temperature rise is due to heat conduction in the ground, which extends to about 10 m outside the borefield even after the heat exchanger heat injection phase. When the heat extraction phase begins at the beginning of the fifth month, the temperature of the ground immediately beyond the borehole wall reduces to about 4 °C and, therefore, the direction of heat conduction in the ground changes towards the borefield resulting in a temperature drop in the borefield and around it.

Figure 10.4 shows borehole wall temperatures for Borehole 1 and Borehole 4 during a 3-year period of heat storage and removal. It is seen that the temperatures of the borehole walls fall on the same path every year. It can be concluded that for a balanced system where borehole spacing, system spacing, and the heat flow rate per unit area are kept at the recommended values and there is no temperature rise or fall after the first year of system performance, there should not be any accumulation

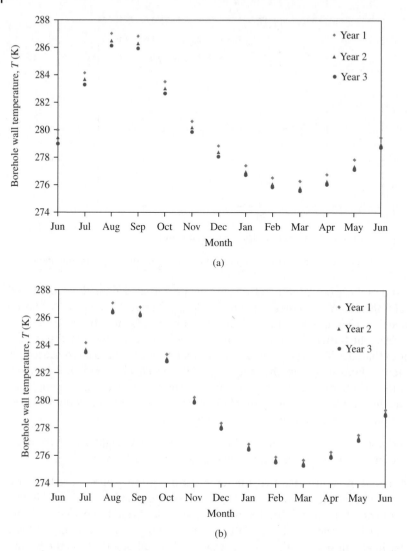

Figure 10.4 Borehole wall temperatures. (a) Borehole 1 and (b) Borehole 4.

of heat in the long term if the same heat storage and removal pattern is followed every year.

However, even the smallest amounts of temperature rise or fall in the ground after the first year of operation can result in unacceptable temperature changes in the long term (Figure 10.5). Figure 10.5 shows the effect of the minor temperature reduction in the ground after the first year of operation over 5 years. It is seen that the temperature at the borehole walls reduces every year and it is estimated that this pattern can affect the heat pump operation over the system's lifetime.

Figure 10.6 shows the temperature contours in the ground surrounding the boreholes at the end of every year in a 5-year period. It is seen that all the heat that is stored in

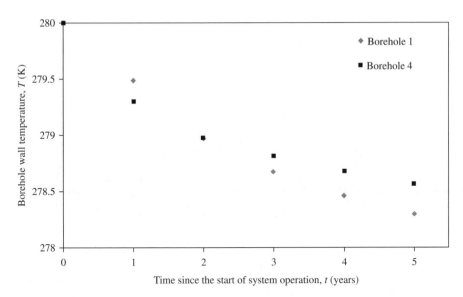

Figure 10.5 Borehole wall temperatures of Borehole 1 and Borehole 4 over 5 years of system operation.

the ground is collected through the GHE. In this case, a small amount of extra heat is also collected which causes a slight temperature reduction in the ground at the end of the first year. This temperature reduction accumulates every year and is doubled by the end of the fifth year of system operation. It is observed that even the slightest amount of excess heat storage or removal, which in some cases can be noticed after the first year of operation, can cause a temperature rise or reduction in the long term and thereby affect the sustainability of the system. However, no significant heat escape is noticed in the case of the current balanced load system.

This effect is also noticed in a recent study by Wang *et al.* (2012). In their study, there was a slightly greater heat extraction from rather than injection into the borehole heat exchanger than required to keep the heat injection and removal in the ground balanced. Consequently, the temperature of the borehole heat exchanger decreased very slightly year on year (0.8 °C after 15 years).

In the example presented in this section, the temperature variations in and surrounding the borefield as a result of a typical closed-loop GSHP operation are shown. Assuming that the borehole heat exchangers are installed in an area that is not close to groundwater or surface waters, the temperature variations that occur in the borefield during its operation may not affect the temperature of surface waters or groundwater. However, small temperature variations are noticed even after the heat injection and extraction cycle of GSHP in regions further away from the borefield. In the presence of surface water flows in such regions, these temperature variations may affect the temperature of the surface water flows in their long-term operation. Although such small temperature variations appear to be minor for most human operations, they may cause disturbances in surface water ecology.

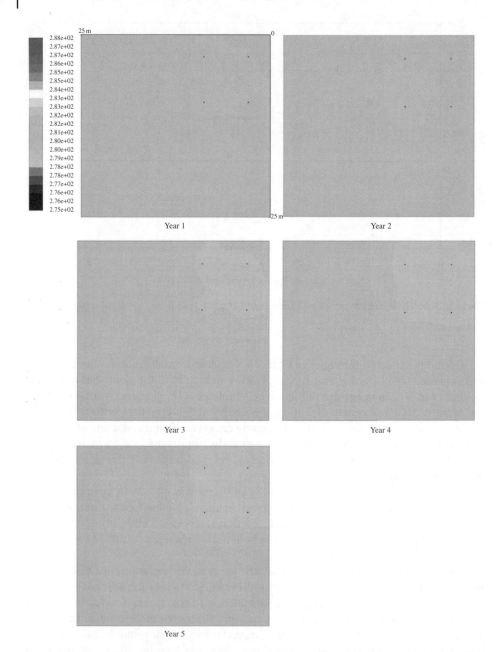

Figure 10.6 Temperature contours in the ground surrounding 4 of the 16 boreholes in the system (the holes surrounded with the highest temperature gradient shown in the figure) for Years 1–5. (Note that four boreholes are shown here due to symmetry.)

References

Andrushuk, R. and Merkel, P. (2009) Performance of Ground Source Heat Pumps in Manitoba, , GeoConneXion Magazine, Summer, pp. 15–16.

Blum, P., Campillo, G., Münch, W. and Kölbel, T. (2010) CO_2 savings of ground source heat pump systems – A regional analysis. *Renewable Energy*, **35**, 122–127.

Fisheries and Oceans Canada (2009) *Fish Habitat and Fluctuating Water Levels on the Great Lakes*, Fisheries and Oceans Canada, Ontario-Great Lakes Area (DFO-OGLA). http://lakehuron.ca/uploads/pdf/fish.habitat.and.GrL.water.levels.fluctuations-DFO .factsheet.pdf (accessed July 27, 2016).

Leong, W.H., Tarnawski, V.R., Koohi-Fayegh, S. and Rosen, M.A. (2011) Ground thermal energy storage for building heating and cooling, in *Energy Storage* (ed. M.A. Rosen), Nova Science Publishers, New York, pp. 421–440.

Markle, J.M. and Schincariol, R.A. (2007) Thermal plume transport from sand and gravel pits – Potential thermal impacts on cool water streams. *Hydrology*, **338**, 174–195.

Sivasakthivel, T., Murugesan, K. and Sahoo, P.K. (2014) A study on energy and CO_2 saving potential of ground source heat pump system in India. *Renewable and Sustainable Energy Reviews*, **32**, 278–293.

Wang, E., Fung, A.S., Qi, C. and Leong, W.H. (2012) Performance prediction of a hybrid solar ground-source heat pump system. *Energy and Buildings*, **47**, 600–611.

York, K., Sarwar Jahangir, Z., Solomon, T., and Stafford, L. (1998) Effects of a large scale geothermal heat pump installation on aquifer microbiota. Proceedings of the Second International Conference on Geothermal Heat Pump Systems, March 16–17, 1998, Richard Stockton College, USA. Pomona.

11

Renewability and Sustainability

11.1 Introduction

In 2013, an estimated 1.2 billion people (17% of the world's population) had no access to electricity (IEA 2015). Although hundreds of millions of people have attained access to modern energy sources through energy programs in the last two decades, lack of energy access still remains one of the main barriers to improving standards of living.

To overcome the many issues associated with energy use, the share of renewable energy sources is expected to increase significantly in the future. Furthermore, technologies are needed that can utilize electricity and other energy forms more efficiently without compromising living standards. The share of worldwide renewable electricity generation has increased to more than 20% worldwide (IEA 2016). To achieve the goal of halving energy related CO_2 emissions of 2005 by 2050 according to the Blue Map scenario set out by the International Energy Agency, current levels of renewable energy use need to be doubled by 2020 and technologies that lead to savings in energy consumption need to be promoted (IEA 2011). Geothermal energy systems are potential options for lowering the electricity use per capita. If they can operate sustainably, ground-source heat pumps (GSHP) are good alternatives to air-source heat pumps, reducing the electricity use significantly and, therefore, mitigate CO_2 emissions and help achieve emissions goals.

11.2 Renewability of Ground-Source Heat Pumps

Geothermal energy systems that utilize hot underground resources and that utilize ambient ground are normally classified as renewable energy technologies. When geothermal energy of either type is utilized, the temperature of the ground is returned either to its elevated temperature by heat contained within hot regions in the earth, or to the background state by the effect of the ambient conditions, which are largely dictated by solar energy.

Installations of GSHP systems have increased globally and in many countries for more than a decade. However, the contribution they make to increased use of renewable energy is sometimes not accepted or recognized in a limited manner. Some possible reasons for this lack of recognition are provided by Lund *et al.* (2004):

Geothermal Energy: Sustainable Heating and Cooling Using the Ground, First Edition.
Marc A. Rosen and Seama Koohi-Fayegh.
© 2017 John Wiley & Sons, Ltd. Published 2017 by John Wiley & Sons, Ltd.

- GSHP systems are associated with the provision of heating and cooling, and therefore unlike geothermal systems using hot underground resources for electricity generation, which receive more attention as renewable energy.
- Some researchers and practitioners question the sustainability of using thermal energy from the ground, perhaps drawing a parallel line with the use of fossil fuels extracted from the earth.
- Some harbor the notion that GSHPs are only an energy efficiency technology because their use involves no net gain in energy output. This idea is probably based on thinking about the operation of air-source heat pumps.
- The latter point can be explained by considering past practice and comparing it with present day technology.
- When air-source heat pumps began to come into use (in the 1950s and 1960s), electricity generation was mainly carried out using central fossil fuel power plants that were less than 30% efficient. The common seasonal performance factor (SPF), which is akin to a coefficient of performance (COP), for air-source heat pumps of that era was 1.5–2.5. As a consequence, on delivery to the building, 60% of the energy is extracted from the air for such technology and only 75% of the energy initially used for electricity generation becomes useful heat (see second column in Table 11.1). In other words, although renewable energy from the air was used to provide thermal energy efficiently, no net gain resulted.
- Considering modern technology, we observe that cogeneration and combined cycle power plants can provide electricity at efficiencies greater than 40%. Common GSHPs exhibit SPF values of 3.5 or higher. As a consequence, the "efficiency" is 140% – implying there is an excess of 40% over the original energy consumed in generating the electricity – and the portion of the product energy derived from the ground is 71% (see third column in Table 11.1, where we consider a water-source heat pump). That is, the combination of a modern, efficient water-source heat pump with a modern, efficient power plant permits the liberation of an excess of renewable energy.

Table 11.1 Comparison of old and modern technology combinations for providing heating using heat pumps.

Parameter	Fossil fuel based electricity generation using old technology combined with old air-source heat pump	Fossil fuel based electricity generation using modern technology combined with modern water-source heat pump
Energy efficiency of electricity generation	0.3	0.4
Seasonal performance factor[a]	1.5–2.5	3.5
Directed energy/consumed energy	0.75	1.4
Delivered renewable energy (% of total)	60	71
Excess renewable energy (% of total)[b]	−25	40

a) Similar to coefficient of performance.
b) Negative value implies a deficit of renewable energy.
Source: Based on data from Lund *et al.* (2004).

Note that, in the comparison considered here, we have not considered electricity generated from renewable energy. If that is the case, then all of the delivered energy is derived from renewable energy, and thus considered renewable energy. Lund *et al.* (2004) point out that it is sensible to integrate electricity generation from renewables with GSHPs to expand the use of renewable energy, and that this could be economic even if the technologies are expensive.

11.3   Sustainability of Ground-Source Heat Pumps

Sustainable geothermal energy utilization often refers to how this energy resource is used to meet current energy needs without compromising its future utilization. Estimating long-term response of geothermal energy sources to current utilization and production capacity is important to discuss their contribution to sustainable development. As a renewable energy source, geothermal energy is often seen to have a significant role to play as a contributor to sustainable development and, more broadly, sustainability. The sustainability of GSHPs depends on the type of systems utilizing energy at various parts of the ground (horizontal heat exchangers versus vertical ones) and the type of ground use (heating only versus heating and cooling). When only heating is concerned, the rate of heat extraction may affect the long-term system operation greatly as its renewal by the geothermal natural heat flows needs to be guaranteed. Horizontal and vertical closed-loop ground heat exchangers are buried at either 2–3 m in depth or at 50–200 m, respectively. In heating using a GSHP coupled to a horizontal ground heat exchanger, heat is fully supplied from solar radiation at a certain rate. Vertical ground heat exchangers operate based on the heat flows from deeper layers of the earth. The sustainability of both systems depends directly on their design (e.g., length, pipe spacing, and depth) as well as operating conditions (e.g., heat extraction rate) to ensure that the heat extraction rate is in the same order as the rate of geothermal heat flows at shallow levels. When heat pumps are used for both heating and cooling, the sustainability of the system concerning the source of energy becomes less of an issue. Sustainability of such systems depends mainly on their design and heat injection and extraction rates from the ground. Properly designed heat pumps extracting and injecting heat at balanced rates and coupled to properly sized and spaced ground heat exchangers are considered to be sustainable. Geothermal energy is considered sustainable not just because it is a renewable energy resource, but also because of its other characteristics that support it being sustainable:

- **Affordability**. Many types of geothermal energy can be utilized today economically for heating and cooling as well as electricity generation, in many locations. Geothermal systems that are not viable economically at present may become economically competitive in the future.
- **Availability**. Geothermal energy in the form of high-temperature ground resources is available in numerous regions of the world, especially those with seismic and volcanic activity. Geothermal energy in the form of ambient ground is available almost everywhere, although its temperature is spatially dependent and affected by the local climate. Note that intermittent renewable energy forms (e.g., solar and wind) exhibit different characteristics to geothermal energy resources.

- **Acceptability**. People are usually supportive of geothermal energy for many reasons, including that it is in many cases renewable, economically viable, non-intrusive, and not visible. The other common renewable energy types, including solar and wind, do not exhibit these characteristics.

11.3.1 Thermal Interaction between Ground-Source Heat Pumps

Although geothermal systems are considered sustainable from several aspects, there are concerns that continued geothermal system development may result in undesirable effects on groundwater resources. The sustainability of GSHP systems at their design efficiency is now being questioned due to "thermal pollution" from the system itself, from adjacent systems, or from the urban environment. Studies from Manitoba, Canada, where the carbonate rock aquifer beneath Winnipeg has been exploited in thermal applications since 1965, indicate that in many cases these systems are not sustainable or not maintainable at their design efficiencies (Ferguson and Woodbury 2005, 2006; Younger 2008). In an area of the carbonate rock aquifer beneath Winnipeg, there are four systems that utilize groundwater for cooling purposes that are closely spaced. Temperatures at the production well have risen as a result of breakthrough of injected water. The results of numerical modeling also indicate that interference effects are present in three of the four systems examined (Ferguson and Woodbury 2006). The influence of these systems on each other implies that these systems have a spacing that is smaller than necessary for such systems, and indicates that there is a limit to the density of development that can occur in a given aquifer. Cases of thermal breakthrough in the aquifer have occurred in some geothermal systems.

In heating- or cooling-dominated climates, an annual energy imbalance is placed on the ground loop due to heating, cooling, and hot water production. In Manitoba's heating-dominated climate, thermal imbalances in the form of large temperature rises in the region where the loop is installed, as well as the migration of thermal plumes away from these systems, can cause significant problems with a heat pump's long-term performance and sustainability if not properly considered at the design phase (Andrushuk and Merkel 2009).

Thermal disturbances in the ground associated with ground heat exchangers are likely to extend beyond property boundaries and affect adjacent properties. Therefore, with increasing interest in installing such systems in the ground and their potential dense population in coming years, procedures and regulations need to be implemented to prevent disputes between neighbors due to potentially interacting systems and their possible negative effects on the design performance of existing nearby systems. As stated by Ferguson (2009), an analogy exists between groundwater and heat flow in the ground which allows one to draw on experiences in groundwater resource development and source water protection. In many ways, the problem of distributing subsurface energy rights is similar to water rights. This argument can even be extended to solar energy systems and disagreements that arise about solar energy rights.

Careful management of geothermal developments to ensure fair access to the subsurface for thermal applications is needed. This will require a greater understanding of subsurface heat flow and input from the scientific and technical communities. These concerns have not been well addressed in all cases. Research is needed to allow the investigation of system performance in an integrated manner, so that the best way of utilizing geothermal systems in a sustainable manner can be determined.

The transient conduction of heat in the ground surrounding ground heat exchangers needs to be modeled in order to evaluate the temperature rise and the heat flow in the ground surrounding boreholes. A preliminary sensitivity analysis of the parameters that affect temperature rise and heat flows in the ground surrounding the boreholes is provided in Appendix B. The existence of thermal interaction among multiple boreholes and their possible negative effects on the design performance of existing nearby boreholes can be examined by a model that accounts for changes in the heat pump COP. Examples of such models were described in Sections 8.1 and 8.2, and a domain consisting of two vertical borehole heat exchangers having a distance of D_b from each other was considered (Figure 8.1a and b). The borehole fluid runs through a U-tube (Figure 8.1c) and delivers or removes heat through the grout in the borehole to the ground surrounding the borehole.

In GSHPs, when cooling the building (ground heat delivery), the borehole fluid temperature can be considered the high-temperature medium and the cooling coil temperature the low-temperature medium in the cooling season. In the heating season (ground heat removal), the heating coil temperature can be considered to be the high-temperature medium while the temperature of the fluid running through the heat exchanger can be considered to be the low-temperature one (see Section 5.1).

In this section, we investigate the effect of thermal interaction on the performance of the heat pump. To simplify, since we are not focused on evaluating the heat pump COP, the reversible (or ideal) heat pump coefficient of performance (COP_{rev}) is used. We now estimate how much the heat pump efficiency varies with a temperature rise of a certain degree in the ground surrounding the borehole. In the presence of specific heat pump data, case specific results for the heat pump COP can be achieved using the same procedure as that used in this study.

In the cooling season (ground heat delivery), the thermal interaction in the ground from a neighboring system can appear in the form of an unwanted temperature rise in the ground surrounding the borehole affecting the borehole fluid temperature and thus the COP. In the heating season (ground heat removal), when the heat is being extracted from the ground, thermal interaction can appear in the form of the escape of heat stored in the ground towards a neighboring operating system that has created a low-temperature region surrounding it by extracting heat from the ground. This may result in lower borehole fluid temperature and lower COP.

Figure 11.1 illustrates the variation in COP_{rev} with borehole fluid temperature for heat delivery and removal modes based on analytical relations. As the system operates in the heat delivery mode, the temperature of the ground surrounding the borehole and, consequently, the temperature of the borehole fluid both increase over time. In the heat removal mode, the temperature of the ground and, therefore, the borehole fluid both decrease as the system operates. It is seen that for various coil temperatures and for both heat delivery and removal modes, the COP decreases as the system operates over time when the borehole wall temperature variation is already high. This could mean that the effect of borehole wall temperature change due to thermal interaction, if there is any, on the performance of the heat pump becomes small if the temperature rise occurs when the borehole wall temperature is already high (in the heat delivery mode) or low (in the heat removal mode).

Figure 11.1 shows the variation in COP_{rev} of the current system, for the geometric specifications in Table 8.1, with the periodic variations of the borehole fluid temperature

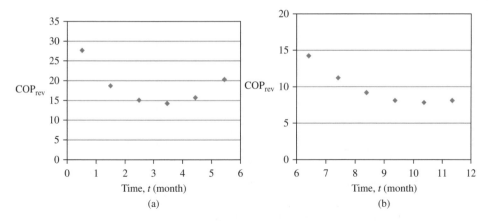

Figure 11.1 Heat pump COP$_{rev}$ variations with time for (a) ground heating load and (b) ground cooling load.

that is shown in Figure 8.20. It is seen that in the cooling season (ground heat delivery), the operation of the heat pump causes the temperature of the borehole wall and, consequently, the temperature of the borehole fluid to increase. An increase in the borehole fluid temperature results in a drop in the heat pump COP as the heat pump has to deliver the heat to an environment with a higher temperature [Equation (5.1)]. In the heating season (ground heat removal), the system operation results in a temperature drop in the ground surrounding the borehole and, consequently, a drop in the borehole wall and the borehole fluid temperatures. The coefficient of performance of the heat pump is lower when it collects heat from a lower-temperature environment, which is the lower temperature of the borehole fluid here [Equation (5.2)].

In the current case, with the geometric specifications in Table 8.1, a large thermal interaction does not occur between the two systems, for the ground heat load given in Figure 8.11. However, if the boreholes are installed closer to each other, thermal interaction is noticed. For example, if the borehole spacing is decreased from 10 m in the current model to 6 m or 4 m, thermal interaction occurs between the boreholes in the form of a temperature rise/drop of less than 0.3 K and 0.6 K on the borehole wall, respectively, due to the operation of the other system. The temperature rise due to thermal interaction on the borehole wall can be evaluated in the current model and is shown for different borehole distances in Figure 11.2. It is seen that as the system experiences a periodic profile for the ground heating and cooling load (Figure 8.11), the temperature of the borehole wall also experiences a periodic profile with a time lag and the temperature rise due to a neighboring system also exhibits a periodic profile. The maximum temperature rise due to a neighboring system depends on the distance between the two systems. The closer the systems, the higher the temperature rise. In addition, the maximum temperature rise due to a neighboring system occurs with a delay after the neighboring system experiences its peak heat load due to the thermal capacitance of the ground. If both systems operate with similar heat load profiles, the system experiences its peak temperature rise due to the neighboring system with a delay after it experiences its peak borehole wall temperature. Depending on the distance between the two systems, this delay may vary. If the boreholes are installed relatively close to each other, this delay will be shorter and the maximum temperature rise due to the neighboring systems occurs

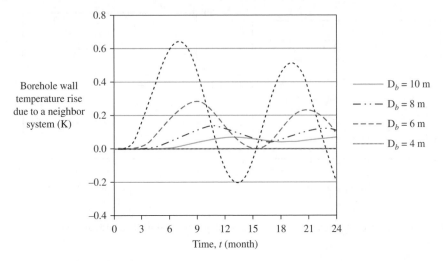

Figure 11.2 Borehole wall temperature rise due to operation of a neighboring system for several values of borehole separation distances D_b.

shortly after the maximum borehole temperature rise is attained on the borehole wall due to the system itself. In this case, the temperature rise that the system experiences due to the operation of another system occurs when the borehole wall temperature and the borehole fluid temperature are already high, and even close to their maxima.

From Figure 11.3a, it is expected that the decrease in COP$_{rev}$ due to this temperature rise will not be great (e.g., a decline of less than 4% in COP in the 295–300 K borehole fluid temperature range). However, when the boreholes are installed further apart, it takes longer until the temperature rise due to maximum ground heat input from the neighboring system reaches the system. When this occurs, the neighboring system may not be at its maximum borehole wall temperature and borehole fluid temperature. In that case, from Figure 11.3a it is expected that the system experiences a larger COP$_{rev}$ decrease due to this temperature rise (e.g., drop of about 10% in COP in the 282–285 K range).

A similar argument can be made when heat is being removed from the ground. In this case, thermal interaction can be interpreted as the rise or drop in the borehole wall temperature and, consequently, in the borehole fluid temperature. In the current system, with characteristics as in Table 8.1, the system experiences a small temperature rise in Months 6–12 due to the neighboring system. This results in a minor increase in the borehole fluid temperature which can actually increase the performance of the heat pump. This is also the case for boreholes that are installed 8 m and 6 m apart. Conversely, it is seen in Figure 11.2 for a borehole distance of 4 m that the system experiences a minor temperature drop in Month 12 when the system is in heat removal mode. A temperature decrease reduces the performance of the heat pump. In this case, at a borehole fluid temperature of 278.9 K, the heat pump COP drops by 3%.

In summary, it can be deduced from Figure 11.2 and Figure 11.3 that the possibility of thermal interaction between two neighboring systems exists when systems are installed relatively close to each other. It is estimated, however, that the thermal interaction between systems that are installed close together will not be large enough to cause

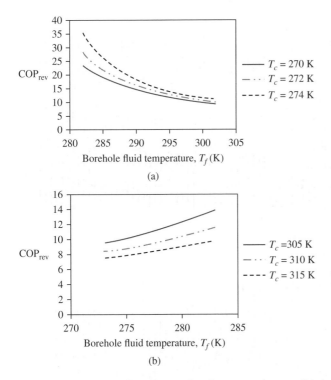

Figure 11.3 Variation of coefficient of performance of a reversible heat pump with borehole fluid temperature in (a) heat delivery mode and (b) heat removal mode for several values of coil temperatures T_c.

decreases in COP_{rev} of more than 10%. Furthermore, the thermal interaction between systems and its effect on COP_{rev} of the neighboring system depends highly on the cycle of the periodic heat input profile of the systems and the distance between borehole systems. This information determines if the temperature rise/drop of a neighboring system reaches the system when it is sensitive to a temperature change or not, and if it actually has a negative effect on the COP_{rev} of the system. For example, the temperature rise in the ground surrounding one system can be transferred towards the other system with a delay of more than 2–3 months and may actually be advantageous to the neighboring system if it is in its heat removal mode. This effect is examined in the longer run over the system's lifetime (30 years). The temperature rise on the borehole wall due to a neighboring system at distance D_b is shown in Figure 11.4. It is seen that the temperature rise oscillates about an average temperature rise that tends to become zero in the second half of the system's operation life. The variation of this average temperature is due to the system reaching a steady-state after the first several years of operation.

It is also seen that, although the oscillations occur between varying temperatures, they occur in the same time periods if the ground heat load is kept constant. For instance, if the temperature rise due to thermal interaction between two boreholes that are 4 m apart occurs in Month 7 of system operation, it is expected to occur repeatedly every year in Month 7. Therefore, a temperature rise noticed in Figure 11.2 in the ground heat removal mode, which is found to be advantageous to heat pump operation, is expected to exist in the same manner during the system's lifetime.

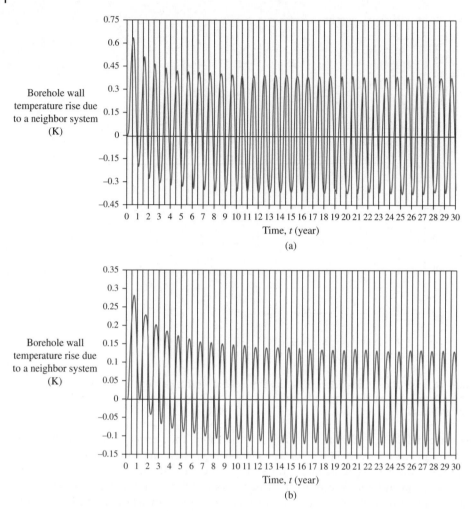

Figure 11.4 Borehole wall temperature rise due to operation of a neighboring system borehole for a separation distance D_b of (a) 4 m and (b) 6 m.

Note that ground heat exchangers are usually installed at system distances greater than 10 m. In larger systems with more than one borehole or with a higher ground heat load, the property area in which the system is being installed is usually large enough to provide adequate distance between the system and its neighbors. However, in cases where the distance between borehole system installations is lower than typical or the system operates with a larger heat load profile, the method presented in the current study can be applied to estimate if there is a thermal interaction between the systems and how it affects the heat pump COP.

References

Andrushuk, R. and Merkel, P. (2009) *Performance of ground Source Heat Pumps in Manitoba*, GeoConneXion Magazine, Summer, pp. 15–16.

Ferguson, G. (2009) Unfinished business in geothermal energy. *Ground Water*, **47** (**2**), 167.

Ferguson, G. and Woodbury, A.D. (2005) Thermal sustainability of groundwater-source cooling in Winnipeg, Manitoba. *Canadian Geotechnical Journal*, **42**, 1290–1301.

Ferguson, G. and Woodbury, A.D. (2006) Observed thermal pollution and post-development simulations of low-temperature geothermal systems in Winnipeg, Canada. *Hydrogeology*, **14** (**7**), 1206–1215.

International Energy Agency (IEA) (2011) Clean Energy Progress Report. https://www.iea.org/publications/freepublications/publication/CEM_Progress_Report.pdf (accessed March 14, 2016).

International Energy Agency (IEA) (2015) World Energy Output 2015 Electricity Access Database. http://www.worldenergyoutlook.org/resources/energydevelopment/energyaccessdatabase/ (accessed March 14, 2016).

International Energy Agency (IEA) (2016) Renewable Energy. http://www.iea.org/aboutus/faqs/renewableenergy/ (accessed March 14, 2016).

Lund, J., Sanner, B., Rybach, L. *et al.* (2004) *Geothermal (ground-source) heat pumps: A world overview*, GHC Bulletin, September, pp. 1–10.

Younger, P.L. (2008) Ground-coupled heating-cooling systems in urban areas: How sustainable are they? *Bulletin of Science, Technology & Society*, **28** (**2**), 174–182.

12

Case Studies

12.1 Introduction

Geothermal energy systems that provide heating and cooling using the ground have become increasingly common. Such systems principally include underground thermal storage, ground-source heat pumps (GSHPs), and district energy. With the introduction of commercial air conditioning equipment in the early 1900s, potential applications increased. The oil embargos of the 1970s suddenly altered energy prices and increased markedly the options for geothermal energy systems that use the ground to provide heating and cooling. Many policies were developed that encouraged the use of GSHPs, thermal energy storage (TES), and district energy to shift electrical demands to off-peak periods and to facilitate the use of renewable energy. As a consequence, geothermal energy systems are utilized today for heating and/or cooling in a wide range of applications.

Geothermal energy systems can be installed and operated cost-effectively in residential, institutional and commercial buildings. Many systems have payback periods of less than 6 years. If time-of-use electricity rates and other financial incentives exist, as is the case for many systems in many jurisdictions, payback periods can be lower. For instance, many utility programs promote energy storage technologies and other demand-side management programs, influencing greatly the economic feasibility of geothermal energy systems. To be effective technically and economically, a geothermal energy system should be designed, sized and operated optimally.

The benefits from using geothermal energy systems for heating and cooling, and their component technologies (underground thermal storage, GSHPs, and district energy), are often numerous, enhancing the attractiveness of the technologies. The benefits normally accrue not only to building owners but also to energy utilities and society.

In this chapter, a range of case studies are presented to illustrate the application of geothermal energy systems that utilize the ground for heating and cooling as well as their advantages and disadvantages. The cases consider applications from the residential, commercial and institutional building sector, as well as relevant utility sector entities involved in electricity generation and district heating and cooling. The case studies illustrate the context in which geothermal energy systems can be employed and assessed, and

Geothermal Energy: Sustainable Heating and Cooling Using the Ground, First Edition.
Marc A. Rosen and Seama Koohi-Fayegh.
© 2017 John Wiley & Sons, Ltd. Published 2017 by John Wiley & Sons, Ltd.

are based mainly on actual applications and drawn from various sources. The types of geothermal energy systems covered through the case studies are as follows:

- underground TES
- ground and water tank TES for heating
- space conditioning with heat pump and seasonal thermal storage
- an integrated system with GSHP, thermal storage, and district energy

12.2 Thermal Energy Storage in Ground for Heating and Cooling

Kumamoto University and Kyushu Electric Power Co. jointly developed a system for storing electricity obtained inexpensively at night as thermal energy in an underground thermal storage, and using it for space conditioning in the day. The daytime is when electrical power grids are pushed to near capacity. The system helps address the daily and seasonal fluctuations in electrical power consumption, which are typically problematic for electricity suppliers, and also leads to reduced emissions of CO_2 and other wastes. The system is briefly described here and further information can be found in the literature (Anon. 1998).

12.2.1 System Description

Soil is the heat storage medium in the underground TES system, which is integrated with the space conditioning system (Figure 12.1).

In the underground TES, 4800 m of flexible 25-mm-diameter plastic water pipe are buried up to 1 m beneath the ground surface. The piping is configured in four layers.

Figure 12.1 The network of water pipes with thermal energy storage for heating and cooling buildings (Anon. 1998).

The TES system occupies about 200 m² (the same area as the building space it is used to heat and cool). A wall of thermal insulation surrounds the ground to the depth below which temperatures are relatively stable throughout the year. This depth is 1.8 m for the present application. The use of the wall of thermal insulation offsets the requirement for an insulating layer at the bottom of the TES.

The system was installed in 1997.

12.2.2   System Operation

Using electricity at night, when rates are low, the system adjusts the temperature of the water running through the pipe. In the summer the system cools the surrounding soil to 10 °C while in winter it heats the surrounding soil to 45 °C. Correspondingly, the stored thermal energy is used for cooling in summer or heating in winter, normally during periods of high electricity usage.

12.2.3   System Advantages

Switching the periods of electrical power use from day (08:00–22:00) tonight (22:00–08:00) under a discounted pricing scheme, by using TES, reduces the electricity bills to 25–35% of what would conventionally be paid. This is despite the fact that the system actually uses 20–30% more electricity than would be the case with a standard space conditioning system.

The system occupies no more area than the space to be heated or cooled. In fact, one option for locating the space conditioning system is to install it directly below the building being heated or cooled.

The system has a low installation cost. It uses a relatively simple structure that avoids the need for laying of concrete and elaborate construction methods.

Using common soil as the TES medium reduces not only daytime electricity consumption but also installation costs. The factors that make a new thermal storage space conditioning system that uses soils as its storage medium advantageous are described in the next subsection. The low construction costs and external space needs of the present system help increase off-peak power consumption by smaller buildings and provide more efficient energy use in Japan.

Besides the fact that customers are charged lower power costs when the system is used, there are other benefits. The soil-based thermal storage system, by transferring some demand to night-time hours, helps level the daytime peaks in electricity consumption and provides constant usage levels. This outcome helps:

- improve the efficiency of existing power plants
- reduce the need for new capacity construction

12.2.4   System Disadvantages

Such systems cost between 20% and 30% more to install than conventional space conditioning systems. This raises the cost relative to more conventional systems for space conditioning.

Also, although these systems may have grown smaller over time as technology has evolved, they often require space on a building's roof for installation.

These systems typically are less attractive for small- to mid-sized buildings where economies of scale are not found.

12.3 Underground and Water Tank Thermal Energy Storage for Heating

A complex and interesting integrated TES system for heating using solar thermal energy is employed in Canada at the Drake Landing Solar Community (DLSC) (McClenahan *et al.* 2006; Wong *et al.* 2006; Sibbitt *et al.* 2007). The thermal storage system consists of an underground thermal storage in the form of a borehole TES, which acts as long-term (seasonal) storage, as well as two water-filled thermal storage tanks which act as short-term storage. The two storage types are integrated and complement one another. The energy system at DLSC combines numerous technologies including thermal storage and renewable energy components to provide more sustainable options that reduce energy costs and environmental impacts.

12.3.1 Location

The DLSC is situated in Okotoks, Alberta, Canada. The town is located 25 km south of Calgary, and has a population of about 28 000 (as of 2015).

12.3.2 Description of the Drake Landing Solar Community

DLSC consists of 52 low-rise, detached houses, which are located within a larger subdivision containing 835 houses. The DLSC houses are located on two streets running east–west. Construction of the DLSC was finished in 2006.

The size of typical DLSC houses, based on gross floor area, is 138–151 m². Each house has a detached garage at the back facing a lane. The garages are joined by a roofed breezeway. The roof structure extends the full length of each of the four laneways.

The DLSC houses incorporate upgraded insulation and heat recovery ventilation. They are certified to the R-2000 standard of Natural Resources Canada. Each house has a low-temperature air handler that blows air across a warm fan coil. After heating, the air is transferred throughout the house through air ducts. The integrated air handler and heat recovery ventilator is composed of a water–air heat exchanger that provides forced-air heating and fresh air.

The primary solar thermal collectors that provide the DLSC houses with space heating are located on top of the roof structure of the breezeway joining the garages. Secondary solar collectors are located on the houses.

12.3.3 Community Energy System

The DLSC houses are supplied with space heating and hot water via rooftop-mounted solar thermal collectors. To facilitate the operation of the DLSC energy system, TES and district heating technologies are also utilized. The DLSC energy system thus forms a community energy system.

The energy system of the DLSC (Figure 12.2) is composed of several principal components. These include two sets of solar thermal energy collectors:

- **Solar thermal collectors for space heating**: Approximately 800 single-glazed flat-plate solar panels are located on the garages of the DLSC houses. There are four rows of garages with two rows of collectors per garage. The solar collectors have an

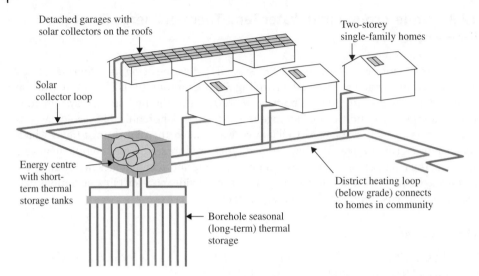

Figure 12.2 Drake Landing Solar Community and its principal energy components. Source: Sibbitt *et al.* (2007). Reproduced with the permission of the Minister of Natural Resources Canada.

area of nearly 2300 m². The solar collectors are linked via an insulated underground pipe carrying glycol (a water and antifreeze solution) to a central Energy Centre.

- **Solar thermal collectors for domestic hot water**: Hot water is produced by a self-regulated, separate solar domestic hot water system using rooftop solar panels. This two-collector solar energy system is separate from the one used for space heating. Natural gas-based hot-water units supplement hot water demands when solar energy is not available and act as back-ups.

One of the rows of houses and garages covered with both sets of solar collectors is shown in Figure 12.3.

Figure 12.3 Row of Drake Landing Solar Community houses and garages covered with solar collectors. Source: Sibbitt *et al.* (2007). Reproduced with the permission of the Minister of Natural Resources Canada.

Figure 12.4 One of the two water-filled, short-term thermal storage tanks at the Drake Landing Solar Community, shown during construction of the Energy Centre in which it is located. Source: Sibbitt *et al.* (2007). Reproduced with the permission of the Minister of Natural Resources Canada.

Additionally, the DLSC energy system has two distinct types of TES:

- **Underground TES.** The underground borehole thermal energy storage (BTES) is located under the corner of a neighborhood park. It consists of 144 boreholes that are 35 m deep and is a long-term (seasonal) storage. The BTES is covered with a layer of insulation and topsoil. Solar-heated water is pumped to the center of the BTES, heating the surrounding ground and permitting the TES to store large quantities of solar heat collected in summer for winter use.
- **Above ground thermal storage tanks.** Two heat storage steel tanks are located in the Energy Centre. Each tank has a capacity of 125 000 l and holds 120 m³ of water, the thermal storage medium. These tanks are used for short-term thermal storage. One of the two water-filled thermal storage tanks is shown in Figure 12.4.

Finally, the DLSC energy system incorporates two other key systems:

- **District heating system.** Heated water is transported from the Energy Centre to the DLSC houses via a district heating loop made up of insulated underground piping. At each house the heated water passes through an air handler in the basement, warming air that is then distributed throughout the house via ducting. The DLSC energy distribution system is laid out as in Figure 12.5. The piping used in the system is shown in Figure 12.6 and the trenches built to accommodate the piping are shown in Figure 12.7.
- **Energy Centre.** Besides the short-term heat storage tanks, the Energy Centre, a 232 m² (2500 ft²) building, contains a back-up gas boiler for peaking requirements during winter and most of the mechanical and electrical equipment (pumps, heat exchangers, controls, etc.). The solar collector loop, the district heating loop, and the BTES loop interface with the Energy Centre.

Pumps for the solar collector and district heating loops use variable speed drives. These allow varied thermal power levels to be accommodated while achieving lower electrical power consumption than conventional drives.

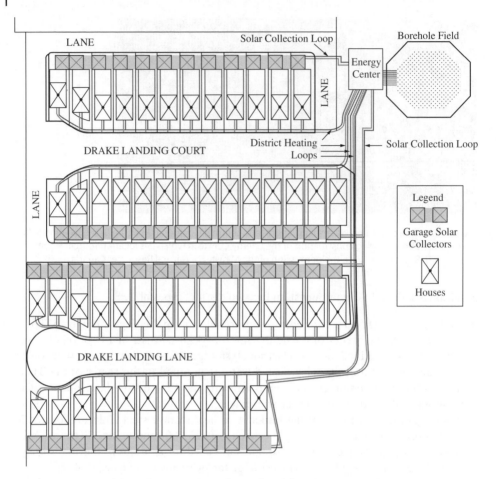

Figure 12.5 Layout of the 52 houses in the Drake Landing Solar Community and its energy distribution system. Source: Sibbitt *et al.* (2007). Reproduced with the permission of the Minister of Natural Resources Canada.

12.3.4   Operation of Energy System

In the Energy Centre, a heat exchanger transfers thermal energy from the solar collector loop to the steel thermal storage tanks, which act as short-term storages (Figure 12.8). Hot water from the tanks can be transferred directly into a district heating network connecting the DLSC houses. During warmer months, hot water from the tanks is circulated to a series of pipes located within boreholes beside the Energy Centre. The BTES stores thermal energy collected in the spring and summer for subsequent use in winter.

The 144 boreholes in the BTES are linked in 24 parallel circuits, each with six boreholes in series. The six boreholes in series are arranged in a radial pattern (Figure 12.9 and Figure 12.10). Water flows from the center to the outer edge when charging the BTES with heat, and from the edge towards the center when recovering heat, maintaining the highest temperature near the center. Heat from the pipes is transferred through a series of U-shaped pipes (Figure 12.11) to the surrounding earth, raising its temperature.

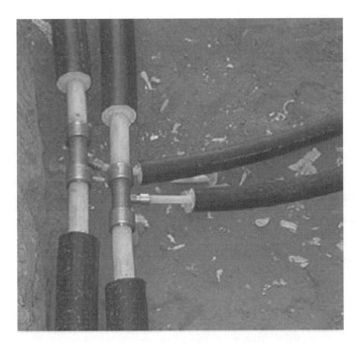

Figure 12.6 Piping used in the Drake Landing Solar Community energy distribution system. Source: Sibbitt *et al.* (2007). Reproduced with the permission of the Minister of Natural Resources Canada.

Figure 12.7 Trenches used for the piping used in the Drake Landing Solar Community energy distribution system. Source: Sibbitt *et al.* (2007). Reproduced with the permission of the Minister of Natural Resources Canada.

The short-term TES tanks act as a buffer between the collector loop, the district energy loop, and the BTES field, receiving and discharging thermal energy as necessary. The short-term TES tanks support the system operation by being able to receive and discharge heat at a much greater rate than the BTES storage, which has a much higher thermal storage capacity.

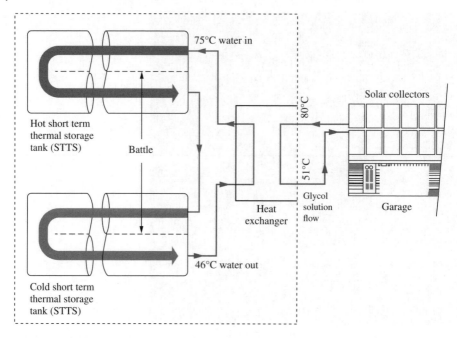

Figure 12.8 Main components of the Drake Landing Solar Community Energy Centre (heat exchanger, the solar collector loop, two thermal storage tanks), with typical temperatures shown for heat transport fluids. Source: Sibbitt *et al.* (2007). Reproduced with the permission of the Minister of Natural Resources Canada.

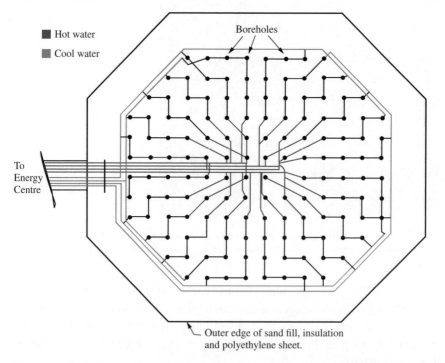

Figure 12.9 Radial layout of the 144 boreholes in the borehole thermal energy storage at the Drake Landing Solar Community in 24 parallel circuits, each with six boreholes in series. Source: Sibbitt *et al.* (2007). Reproduced with the permission of the Minister of Natural Resources Canada.

Figure 12.10 Borehole thermal energy storage construction at the Drake Landing Solar Community, showing radial configuration of boreholes and the piping. Source: Sibbitt *et al.* (2007). Reproduced with the permission of the Minister of Natural Resources Canada.

Figure 12.11 U-shaped pipes that descend in each 35-m-deep borehole at the DLSC. Source: Sibbitt *et al.* (2007). Reproduced with the permission of the Minister of Natural Resources Canada.

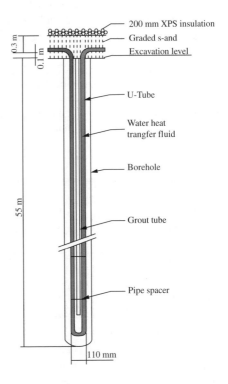

The interaction between the long- and short-term storages is significant in various situations. For example, during periods of high solar insolation, the BTES cannot receive energy as quickly as it can be collected, so it is temporarily stored in the short-term TES tanks and subsequently transferred to the BTES at night. Also, when heat cannot be discharged from the BTES sufficiently quickly to meet winter peak heating demands, usually in early mornings, heat is continually removed from the BTES and stored in the short-term storage tanks until needed.

12.3.5 Technical Performance

For the houses in the DLSC, 90% of heating and 60% of hot water needs are designed to be met using solar energy. This is achieved via the approximately 800 single-glazed flat-plate solar panels organized in four rows on the garages, which can generate 1.5 MW of thermal power on a typical summer day. This energy is stored until needed for space heating.

The annual heat demand per DLSC house is 50 GJ. Annually, each house uses approximately 110.8 GJ less energy than a conventional Canadian house (Table 12.1) and emits about 5.65 t fewer greenhouse gases (Table 12.2). Thus, the DLSC avoids about 260 t of greenhouse gas emissions annually. Each DLSC house is about 30% more efficient than conventionally built houses and is expected to use 65–70% less natural gas to heat water than a conventional new house.

The solar thermal energy transferred to the BTES raises the temperature of the surrounding soil and rock to about 80 °C by the end of summer. During winter, water at approximately 40–50 °C from the boreholes is transferred by the heat exchanger to the district energy network for circulation to houses. But, since large seasonal TES systems require a significant time to charge since the storage medium must be heated up to a

Table 12.1 Comparison of annual energy use (in GJ) for a Drake Landing Solar Community house and a conventional house, for space and domestic hot water heating.

Energy service and source	Conventional house	Drake Landing Solar Community house	Reduction (relative to conventional house)[a]	
			Absolute reduction	Percentage reduction
Space heating				
Natural gas	100	6	94	94
Solar energy	0	62	−62	—
Total	100	68	32	32
Water heating				
Natural gas	26	9	17	65
Solar energy	0	9	−9	—
Total	26	18	8	69
Space and water heating				
Natural gas	126	15	111	88
Solar energy	0	71	−71	—
Total	126	86	40	68

a) The reductions in rows labeled "Natural gas" reflect reductions in natural gas use for a DLSC house with respect to a conventional house. The reductions in rows labeled "Total" reflect the reduction in energy required and thus reflect savings due to conservation measures in the DLSC houses. The negative absolute reductions in rows labeled "Solar energy" reflect the fact that Drake Landing Solar Community homes use solar energy and conventional homes do not.

Table 12.2 Comparison of annual greenhouse gas emissions (in t) for a Drake Landing Solar Community house and a conventional house, for space and domestic hot water heating.

Energy service	Conventional house	Drake Landing Solar Community house	Reduction (relative to conventional house)	
			Absolute reduction	Percentage reduction
Space heating	5.1	0.3	4.8	94
Water heating	1.3	0.5	0.9	69
Space and water heating	6.4	0.8	5.7	89

minimum temperature before any heat can be extracted, it is anticipated that by the fifth year of operation the system will reach a 90% solar fraction, defined as the ratio of the amount of energy provided by the solar technologies to the total energy required.

The thermal capacity of the district heating network is 4.5 MW.

12.3.6 Economic Performance

DLSC house owners receive a monthly solar utility bill for heating of $60 on average. The DLSC houses sold for an average of $380 000. All monetary values in this subsection are in Canadian dollars.

The initial start-up capital for the Drake Landing system was $7 million. This included financial incentives of $2 million from federal government agencies, $2.9 million from the Green Municipal Investment Fund of the Federation of Canadian Municipalities and $625 000 from Innovation Program agencies of the government of Alberta.

The initial capital cost of such a project in the future is estimated to be $4 million. This is $3 million less than the actual $7 million project cost because that included numerous one-time research and development expenses that would not be necessary if the project were replicated.

The use of a low-temperature hot water system permitted the capital cost of piping to be reduced.

The optimal size for such a community has been estimated based on economies of scale and data for 2008 to be a minimum of 200–300 houses. The system would be the same except that more boreholes would be required.

The DLSC project demonstrates the feasibility of replacing substantial residential conventional fuel energy use with solar energy, collected during the summer and utilized for space heating during winter, in conjunction with seasonal thermal energy.

12.4   Space Conditioning with Heat Pump and Seasonal Thermal Storage

A space conditioning system incorporating a heat pump with underground seasonal thermal storage used in the form of aquifer TES is used by the Anova Verzekering Co. Building in Amersfoort, The Netherlands. The system was installed in an office building when it was renovated in the early 1990s (IEA-HPC 1994).

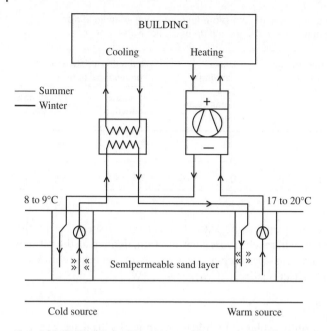

Figure 12.12 Space conditioning system incorporating a heat pump with underground seasonal thermal storage using in the form of aquifer thermal energy storage, showing heating and cooling modes. Source: Adapted from IEA-HPC (1994).

12.4.1 Description of System

A schematic diagram of the system for both heating and cooling modes of operation is shown in Figure 12.12. The system includes an electric heat pump, which is mainly used in winter, as well as a heat exchanger, which is mainly used in summer.

A hydronic heating and cooling distribution system is also incorporated into the system. The hydronic heating and cooling distribution system is ceiling mounted in the design.

12.4.2 System Operation

The system has two main operating modes, based on season:

- **Heating mode.** In winter, the system operates in the heating mode. Then, the heat pump supplies the hydronic heating and cooling distribution system with heat extracted from an underground heat source in the form of warm groundwater from an aquifer. The cooled groundwater is stored in a second water layer location.
- **Cooling mode.** In summer, the system operates in a cooling mode. When cooling is needed, the cooled groundwater is pumped back to the surface and used as a cooling medium to extract heat via the hydronic system. The spent groundwater is returned to the warm well and stored there.

12.4.3 Technical Performance

The heat pump provides a warm fluid to the heat-distribution system at a temperature normally between 35 °C and 40 °C.

Table 12.3 Comparison of annual energy use for a heat pump system with underground seasonal thermal energy storage and a conventional system.

Energy type	Annual energy use		Reduction in annual energy use relative to conventional system	
	Conventional system	Heat pump and underground seasonal thermal energy storage-based system	Absolute	Relative (%)
Natural gas (m³)	215 800	95 500	120 300	56
Electricity (kWh)	395 550	511 500	−84 000	−21
Primary energy (m³)	322 000	179 000	143 000	44

Source: Adapted from IEA-HPC (1994).

The cooled groundwater that is stored in a second water layer location is typically at a temperature of 8–9 °C. Enough of this cool water is stored by the end of the heating season to satisfy the office cooling needs throughout the summer without using an active cooling system operation or the heat pump in the cooling mode.

The spent groundwater that is returned to the warm well is typically at a temperature of 17–20 °C. This temperature is sufficiently high to permit the stored heat source to be used as a heat source for winter heating.

This system achieves significant net energy use reductions, as seen in Table 12.3. The electricity use increases since the heat pump is electrically driven. Nonetheless, total primary energy use due to the utilization of the heat pump and aquifer TES system is observed to be lowered by more than 40%. The total primary energy is the equivalent amount of natural gas, assuming 1 kWh of electricity generation requires 0.25 m³ of natural gas.

The heat pump operation is efficient. This is in large part because the system uses a consistent-temperature warm heat source combined with a consistent-temperature temperature heat distribution medium.

This application shows how an aquifer TES system provides energy savings and reductions in environmental pollutants.

12.4.4 Economic Performance

The total installation cost was over US$1 000 000. A subsidy of US$212 000 was provided by the government of the Netherlands. The subsidy is significant, accounting for more than 20% of the total installation cost.

The reductions in energy costs compared with conventional systems for space heating and cooling provide a payback for the additional investment costs of less than 6.5 years. The conventional systems considered for this assessment are gas heating for space heating and electric air conditioning for space cooling.

12.4.5 Environmental Performance

This system achieves significant reductions in environmental emissions, as shown in Table 12.4. Reductions of 40% or more are observed in emissions of CO_2, nitrogen oxides (NOx) and sulphur dioxide (SO_2).

Table 12.4 Comparison of annual environmental emissions for a heat pump system with underground seasonal thermal energy storage and a conventional system.

Emission type	Annual emission (Mg)		Reduction in annual emission relative to conventional system	
	Conventional system	Heat pump and underground seasonal thermal energy storage-based system	Absolute (Mg)	Relative (%)
CO_2	608	346	262	43
NOx	—	—	—	40
SO_2	—	—	—	40

Source: Adapted from IEA-HPC (1994).

12.5 Integrated System with Ground-Source Heat Pump, Thermal Storage, and District Energy

An integrated system with GSHPs, underground TES in the form of BTES, and local district energy exists at the University of Ontario Institute of Technology (UOIT) in Oshawa, Ontario, Canada. The university has about 10 000 students divided among two campuses. Most of the north campus seven buildings are designed to be heated and cooled using the integrated system, with the aim of reducing energy use, environmental emissions, and economic costs. This system was introduced in Section 4.4.5 and considered in a detailed thermodynamic assessment in Section 8.5.

12.5.1 Nature of the Integration

The UOIT facility is an integrated system. It combines:

- an energy resource (the ground)
- a system to exploit this energy [the heat pump, heating, ventilating, and air conditioning (HVAC) and distribution equipment and related devices]
- an interface (geothermal heat exchangers and an underground storage)

The design of the UOIT system is tailored to the southern Canadian context, which has significant space heating and air conditioning needs. Energy is upgraded by heat pumps for heating, that is, heat is taken from the ground at low temperature and transferred at a higher temperature to the building. Also, the ground can absorb energy and be increased in temperature using the heat pump in its cooling (reverse) mode. Thus, the system provides for both heating and cooling on seasonal bases. The UOIT geothermal well field is the largest and deepest in Canada and one of the largest in North America, and it is unique in Canada in terms of the number of holes, capacity, and surface area.

12.5.2 Local Ground Conditions

The local hydrogeology in the vicinity of the system in Oshawa and its thermal characteristics were assessed with test drilling and in-situ tests (Beatty and Thompson 2004).

12.5.2.1 Geology and Soil

The geology of the ground is heterogeneous. Downward from the surface, the ground contains (see Figure 4.6):

- **44 m of unconsolidated overburden deposits**. The overburden comprises layers of glacial till, clay, silt, and silty fine sand. The sand deposits are not water-bearing.
- **Shale and limestone bedrock**. These two Paleozoic sedimentary bedrock formations are broken down as follows: 14 m of shale overlying 142 m of almost impermeable limestone (encountered 55–200 m below the surface). Few fractures or fissures are present in the limestone.

12.5.2.2 Groundwater

Groundwater resources in the region are limited to isolated, thin sand deposits. The fact that little groundwater flow exists in the homogeneous, non-fractured rock to transport thermal energy from the site renders the ground suitable for thermal storage.

12.5.2.3 Temperature and Thermal Characteristics

The local background temperature of the ground, found far below the surface, is 10 °C. The temperature in the region within a few meters of the surface varies with the local ambient air temperature and conditions.

The thermal conductivity for the ground encountered in a test well was determined with thermal conductivity tests, and found to be approximately 1.9 W/m K (Beatty and Thompson 2004).

12.5.3 Design of Borehole Thermal Energy Storage System

The BTES field occupies the central quadrangle of the university north campus (see Figure 4.10). For the total predicted cooling load of the campus buildings (7000 kW), a field of about 370 boreholes, each about 200 m in depth, was determined to be required to satisfy the energy service needs. Five temperature-monitoring boreholes were also installed. The borehole heat exchangers are located on a 4.5 m grid. The total field approximately has a surface area of 7000 m², a volume of 1.4 million m³, and a mass of 2 300 000 t (1 700 000 t of rock and 600 000 t of overburden).

About 75 km of drilling was required. Correspondingly, approximately 150 km of polypropylene tubing (10 cm diameter) routes the water down the borehole depth and back up to the mechanical corridors that surround the field. Water-filled instead of grouted borehole heat exchangers were utilized, with the aim of improving the efficiency of U-tube installation and extend borehole life. Steel casing was installed in the upper 58 m of each borehole to seal out groundwater in the shallow formations.

The system uses GSHPs to achieve the required temperatures, with a glycol solution circulated in polyethylene tubing through an underground network. Fluid circulating through tubing extended into the wells collects heat from the earth and transports it to buildings in winter, and extracts heat from buildings and transports it to the ground in summer.

The BTES field is divided into four quadrants to permit storage of heat and cold simultaneously, and thereby to assist in achieving optimal seasonal energy storage.

The well field of the system during construction, showing the borehole grid and interconnecting piping, is presented in Figure 12.13. Uncapped well heads can be seen in the figure.

Figure 12.13 View of borefield during the construction of university buildings and the borehole thermal energy storage system, showing the grid of borehole headers (black dots) and interconnecting piping.

Start-up of the system occurred during the summer of 2004. Monitoring of the fluid and energy flow in the first few years of operation has been required to optimize the long-term performance of the BTES system. A string of temperature probes in each of the five monitoring wells monitor the thermal store within and outside the BTES field.

The BTES field construction and borehole heat exchanger installation have been described in detail elsewhere (Beatty and Thompson 2004), and are summarized in the next two subsections.

12.5.4 Borefield Drilling

Before borehole drilling, approximately 2 m of soil was removed to create a level base for the BTES field. A 300-mm-layer of crushed stone was placed over the glacial till soils to provide a working base for the heavy drilling rigs and a drainage layer for precipitation and drilling fluids.

Drilling, at a rate of about 15 m/day, was performed using two drilling techniques to accommodate the site geology. Hydraulic mud rotary drilling was used in the upper 56 m of overburden and shale, while an air-driven, down the hole hammer was used in the limestone.

Steel casings of 150 mm diameter with threaded couplings were seated about 1.5 m into the limestone bedrock of each borehole. To prevent downward seepage of surface drainage, the surface annulus around each casing was sealed with bentonite grout.

The drilling was inspected daily for stratigraphy and groundwater conditions in the boreholes, and to confirm depths. About 10% of the boreholes were checked for plumbness, limestone formation fractures, and groundwater seepage.

At the completion of drilling, the boreholes were left dry and were ready for the installation of the borehole heat exchangers. The water level recovery rate was low, and monitoring of three observation wells found the value to average about 2 cm per day over 200 days.

12.5.5 Borehole Heat Exchanger Installation

In each borehole, a borehole heat exchanger consisting of a polyethylene U-tube through which a heat-transfer fluid circulates was installed. Water-filled boreholes were used in the UOIT system, resulting in several cost-savings over conventional grouted U-tubes. The UOIT high-density polyethylene U-tubes have a 32 mm inner diameter.

During the 370 U-tube installations, 2-cm-diameter iron sinker bars with a mass of 90 kg were attached to the bottom 20 m of the U-tube to counteract the buoyancy of the tubing in the water-filled borehole. As the 200 m long U-tube assembly was inserted into each dry borehole (see Figure 4.6), the U-tube and the borehole were simultaneously filled with treated water from the local water supply system. Fusion caps were installed on the pipe ends to prevent entry of any surface water or debris. The U-tube installation rate was about 6000 m/day. A borehole and its U-tube heat exchanger are illustrated in Figure 12.14.

After installation, each heat exchanger assembly was pressure tested by pressurizing the water in the U-tube with compressed air to at least 690 kPa and maintaining the pressure for 1 h. The U-tube was deemed leak-free if little pressure was lost at the end of the 1-h test period. If U-tubes exhibited excessive pressure loss, they were replaced.

After the heat exchangers were installed and tested, each borehole was topped up with municipal water and an air-tight sanitary well seal was placed on top of the borehole casing. After connection of the heat exchanger tubes to the horizontal distribution pipes, the BTES field was backfilled with about 2 m of clean fill. The casings on the six temperature-monitoring wells and one of the heat exchange boreholes were extended up to the final grade level to permit easy access for future instrumentation and monitoring or other purposes.

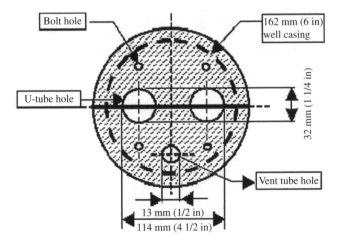

Figure 12.14 Cross section of borehole and U-tube borehole heat exchanger.

12.5.6 Integration of Ground-Source Heat Pumps and Heating, Ventilating, and Air Conditioning

The central plant contains a GSHP system that is integrated with the underground thermal storage. The university buildings are attached to a central plant by a grid of borehole wells. A flow chart for the BTES system is presented in Figure 12.15, and the technical specifications of the chiller and two heat pumps, each having seven modules, are listed in Table 12.5. The integrated system consists of a primary system (with heat pumps, chillers, boilers and the underground thermal storage) and a secondary system (with circulation pumps and air handling units).

12.5.6.1 Primary System

Chillers are used to pump energy from the buildings into the BTES. Heat pumps run only in the chilling mode for cooling. Chilled water is supplied from two chillers, each

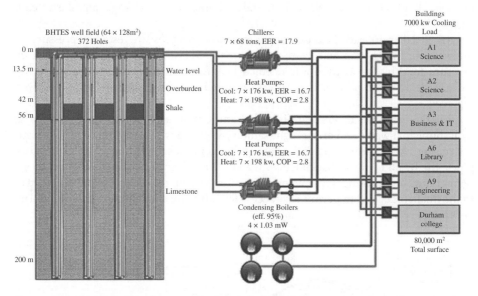

Figure 12.15 Schematic flow diagram of the borehole thermal energy storage at the University of Ontario Institute of Technology.

Table 12.5 Design parameter values for heat pumps.

Design parameter	Operating mode	
	Cooling	Heating
Coefficient of performance	4.9	2.8
Total energy load (kW)	1236	1386
Load water inlet/outlet temperatures (°C)	14.4/5.5	41.3/52
Source water inlet/outlet temperatures (°C)	29.4/35	9.3/5.6

having seven 90-t modules, and two sets of heat pumps with seven 50-t modules each. The 90-t modules are centrifugal units with magnetic bearings that allow for very good part-load performance. The condensing water is transported to the borefield.

The borefield retains the condensing heat for use in the winter (when the heat pumps reverse) and provides low-temperature hot water for the campus. Most services use this 53 °C hydronic heat. Each building is hydronically isolated with a heat exchanger, and has an internal distribution system. Supplemental heating is also provided by condensing boilers.

In autumn, energy is recovered from the borefield, and the return water is hot enough to be used for "free-heating," avoiding heat pump use.

12.5.6.2 Secondary System

The circulation pumps are installed in banks, with rotating duty cycles, in part to ensure continuous availability even in the event of failures. All the motors are controlled by the central control system and have variable frequency, variable speed drives, which provide continuously variable and controllable flow rates. These drives facilitate optimal heat transfer for the chillers and heat pumps.

Air handling units condition, filter and feed the air to building rooms. Air is monitored for temperature, humidity and CO_2. The CO_2 concentration is used to meter the amount of outside fresh air required. This outside air is preheated using heat from the exhaust air.

12.5.7 Techno-Economic Performance

12.5.7.1 Technical Performance

The technical performance of the integrated system is affected by outdoor conditions, the return temperature of the heat distribution network, the characteristics of the thermal energy demand(s), and heating equipment. Decreasing the supply temperature increases the size of the building heating equipment, which is significant since oversizing leads to increased cost and exergy destruction, due to such irreversibilities as pumping and pipe friction.

This technical performance of the integrated system was evaluated via a detailed thermodynamic assessment in Section 8.5.

Additionally, the behavior of the system heat pumps can be assessed by considering three coefficients of performance (COPs). The variation of these three COPs with the exit temperature of the heat pump (or supply temperature of the heat distribution system) in the heating mode is listed in Table 12.6 for a typical set of conditions. There the conventional COP (or "energy efficiency") of the heat pump with a heating load rate $\dot{Q}_h$ and a compressor work input rate $\dot{W}_{comp}$ follows:

$$\text{COP}_{\text{actual}} = \frac{\dot{Q}_h}{\dot{W}_{comp}} \tag{12.1}$$

The Carnot COP represents the maximum heating COP for a heat pump operating between a low-temperature reservoir T_L and high-temperature reservoir T_H:

$$\text{COP}_{\text{Carnot}} = \frac{T_H}{T_H - T_L} \tag{12.2}$$

Table 12.6 Variation of coefficient of performance with heat pump supply temperature

Heat pump supply temperature (°C)	Coefficient of performance		
	Actual	Exergetic	Carnot
45.0	1.7	0.3	6.6
47.5	2.4	0.4	6.4
50.0	3.1	0.5	6.1
52.5	3.4	0.6	5.9
55.0	4.1	0.7	5.6

Based on the above two measures, an exergetic COP (i.e., efficiency ratio) can be written as:

$$\text{COP}_{\text{exergetic}} = \frac{\text{COP}_{\text{actual}}}{\text{COP}_{\text{Carnot}}} \tag{12.3}$$

Table 12.6 indicates that raising the supply temperature increases the exergy efficiency of the heat pumps and hence the overall system. In heating systems, the heat distribution network supply temperature usually affects exergy loss significantly. Determining the appropriate temperature is challenging because increasing it lowers the investment cost for the distribution system and the electrical energy required for pumping, but raises distribution network heat losses.

12.5.7.2 Economic Performance

The integrated system was designed in large part to reduce lifetime economic costs. As is commonly the case with sustainable energy systems, higher initial capital costs (for the geothermal borefield as well as high-efficiency HVAC equipment here) are required to achieve energy cost savings over time, and that is the case for the present integrated system. The economic performance is assessed here by considering the annual energy savings attributable to the use of the integrated system instead of more conventional technology, as well as the payback periods.

Considering only direct financial cost reductions attributable to the integrated system, the following key financial savings parameter values were obtained:

- Annual energy costs for heating are reduced by 40%.
- Annual energy costs for cooling are reduced by 16%.

Also, the following important economic performance parameters were determined (for the system designed) by relating the annual energy cost savings to the overall and component system financial costs:

- The simple payback period for the geothermal well field is 7.5 years.
- The simple payback period for the high-efficiency HVAC equipment is 3–5 years.

Various other indirect financial benefits attributable to the system, although real, were not considered in the above findings. The principal indirect financial benefits include reductions in the avoided costs associated with the boiler plant, rooftop cooling towers (along with related building support), the use of potable water (23 000 000 l annually) and the use of chemicals for water treatment.

Given the integrated system has a lifetime of at least 20 years for the main HVAC components and longer for the BTES, the lifetime economic benefits are significant and appear to make the integrated system meritorious.

12.6 Closed-Loop Geothermal District Energy System

A closed-loop geothermal district energy system, composed of boreholes, energy stations, hot and cold district loops of water-filled pipes, and building interfaces, is installed at Ball State University in Muncie, Indiana, USA, a campus covering about 3.0×10^6 m^2 (731 acres) (Boyd *et al.* 2015). Originally, the university's heating needs were covered by four coal boilers burning 36 000 t of coal per year. The new geothermal system is to be the largest geothermal system installed in the USA.

12.6.1 System Description

A 455 kW (130 t) GSHP (Boyd *et al.* 2015) coupled with a vertical closed-loop piping system uses the ground as either a heat source, when operating in heating mode, or a heat sink, when operating in cooling mode.

The system consists of 3600 12.7-cm-diameter (5-in.) boreholes (also reported to be 4100 boreholes) (Lund *et al.* 2010; Boyd *et al.* 2015) and spans more than 600 acres throughout the campus, under recreational areas, parking lots, and open green space within the campus quadrangle areas. The boreholes are drilled to depths of 122–152 m (400–500 ft) and are connected and routed to three different energy stations located throughout the campus at a depth of 1.5 m (5 ft) below finished grade. The U-tubes placed in the boreholes do not require antifreeze and circulate water.

In each of the energy stations, heat is extracted from or delivered to the ground to hot and cold water loops using a heat pump that uses refrigerant R134A. The hot water is supplied at a temperature of about 66 °C (140 °F) and the chilled water at approximately 7 °C (45 °F). Two separate district loops provide the distribution of water for heating and cooling purposes. Originally, the campus had a district cold water distribution system and the geothermal conversion project includes the addition of a district hot water distribution system throughout the campus. The loops pass through heat exchangers and fans cool or warm 5.1×10^5 m^2 (5.5 million ft^2) of Ball State's 47 buildings across the campus to distribute energy during the heating and cooling cycles of the year based on the thermostat settings of the individual buildings.

The horizontal looping network of supply and return lines from the energy stations to the buildings on the campus also can be used to redistribute energy from one building to another. For a given building that is overheating or underheating, energy can be exported or imported to or from an adjacent facility experiencing the opposite energy need. Natural gas boilers provide peaking steam needs and back-up to the geothermal system. The conversion to geothermal includes the installation of a campus wide new hot water supply and return distribution system.

The installation of the system, which began when the first hole was drilled in the ground in 2009, occurred in two phases. Phase 1 consisted of installation of two borefields containing 1800 boreholes, construction of the North District Energy Station and connecting buildings on the northern part of the campus to the new distribution system. Phase 2 includes construction of the system's south borefield containing 1800 boreholes, modifications to the South District Energy Station to accommodate two 8792-kW (2500-t) heat-pump chillers, hot and chilled water distribution looping, and connecting the remaining buildings (predominately on the south side of campus) to the geothermal systems (Ball State University 2016).

12.6.2 Efficiency

It is estimated that the COP will increase from an equivalent value of 0.62 when using the coal power plants to 7.77 for the district geothermal system (Department of Energy 2010).

12.6.3 Economics

Originally, the university's heating needs were covered by burning coal at a cost of $3.2 million per year. The current total project estimate is about $80 million and it is expected to lower energy costs more than $2 million annually, compared with operating the coal plant. The cost of operating the system is only to power the system's equipment. All monetary values in this subsection are in US dollars.

12.6.4 Environmental Benefits

Annual emissions of tons of nitrous oxide, particulate and SO_2 emissions are eliminated through use of the heat pump system that uses an environmentally friendly refrigerant. The projected reduction in carbon emissions is estimated to be more than 7.3×10^7 (80 000 t) per year. The overall switch to the geothermal system will allow Ball State University to cut its carbon emissions to about half its original amount.

12.7 Closing Remarks

A range of case studies of geothermal energy systems that utilize the ground for heating and cooling are presented. Included are underground TES systems and systems with ground and water tank TES for heating and/or cooling, space conditioning systems with heat pumps and seasonal thermal storage, as well as integrated systems including GSHPs, thermal storage, and district energy. The case studies demonstrate how geothermal energy systems can be designed and operated as a part of the overall energy infrastructure for a building or facility. The applications covered by the case studies include residential, institutional, educational, and commercial facilities. Any geothermal energy system typically has advantages and disadvantages, usually linked to the application and location. The case studies demonstrate the various ways in which geothermal energy systems can be effective, efficient, economic and environmentally benign means for utilizing the ground for heating and cooling. Such systems can thereby contribute notably, now and even more in the future, to energy sustainability in particular and overall sustainability in general.

References

Anon. 1998 Going Underground: New Thermal Storage System to Save Energy. Trends in Japan (Feb. 17). http://web-japan.org/trends98/honbun/ntj980217.html (accessed January 8, 2016).

Ball State University 2016 Going Geothermal. http://cms.bsu.edu/about/geothermal (accessed January 27, 2016).

Beatty, B & Thompson, J 2004 75 km of drilling for thermal energy storage. Geo-Engineering for the Society and its Environment: Proceedings of the 57th Canadian Geotechnical Conference, October 24 – 27, 2004, Quebec City, Canada. Canadian Geotechnical Society, Session 8B, pp. 38-43.

Boyd, TL, Sifford, A, & Lund, JW 2015 The United States of America country update. Proceedings of the World Geothermal Congress 2015, April 19 – 25, 2015, Melbourne, Australia, pp. 1 – 12. http://www.geothermal-energy.org/pdf/IGAstandard/WGC/2015/01009.pdf (accessed January 27, 2016).

Department of Energy (2010) Recovery act state memos, Indiana 2010. http://energy.gov/sites/prod/files/edg/recovery/documents/Recovery_Act_Memo_Indiana(1).pdf (accessed January 27, 2016).

International Energy Agency's Heat Pump Centre (IEA-HPC) (1994) Energy Storage, International Energy Agency. *IEA-Heat Pump Center Newsletter*, **12 (4)**, 8.

Lund, JW, Gawell, K, Boyd, TL, & Jennejohn, D 2010 The United States of America country update. Proceedings of the World Geothermal Congress, April 25 – 29, 2010, Bali, Indonesia, pp. 1 – 18. http://www.geothermal-energy.org/pdf/IGAstandard/WGC/2010/0102.pdf (accessed January 27, 2016).

McClenahan, D., Gusdorf, J., Kokko, J. *et al.* (2006) Okotoks: Seasonal Storage of Solar Energy for Space Heat in a New Community. ACEEE, 2006 Summer Study on Energy Efficiency in Buildings, , Pacific Grove, CA.

Sibbitt, B, Onno, T, McClenahan, D, Thornton, J, Brunger, A, Kokko, J, & Wong, B 2007 The Drake Landing Solar Community project: Early results. Proceedings of the. Canadian Solar Buildings Conference, June 10 – 14, 2007, Calgary, Canada.

Wong, W.P., McClung, J.L., Snijders, A.L. *et al.* (2006) First large-scale solar seasonal borehole thermal energy storage in Canada. Ecostock 2006 Conference, , Stockton, USA.

A

Numerical Discretization

The discretization in unstructured meshes for the grout and the ground can be developed from basic control volume techniques where the integral form of the energy conservation equation is used as the starting point:

$$\int_{CV} \text{div}\,[\text{grad}(T)]\,dV = \int_{CV} \frac{1}{\alpha}\frac{\partial T}{\partial t}dV \tag{A.1}$$

Here, V is the volume. Integration of Equation (A.1) over a time interval from t to $t + \Delta t$ gives:

$$\int_{t}^{t+\Delta t}\int_{CV} \text{div}\,[\text{grad}(T)]\,dV\,dt = \int_{t}^{t+\Delta t}\int_{CV} \frac{1}{\alpha}\frac{\partial T}{\partial t}dV\,dt \tag{A.2}$$

The volume integration on the right-hand side can be conveniently evaluated as the product of the volume of the cell and the relevant centroid value of the integrand. The time integration in the current model is treated using the implicit technique.

Using Gauss's divergence theorem, which is applicable to any shape of control volume:

$$\int_{CV} \text{div}\,\mathbf{a}\,dV = \int_{A} \mathbf{n}\cdot\mathbf{a}\,dA \tag{A.3}$$

The diffusive terms on the left-hand side of Equation (A.2) are rewritten as integrals over the entire bounding surface A:

$$\int_{t}^{t+\Delta t}\int_{A} \mathbf{n}\cdot\text{grad}\,T\,dA\,dt = \int_{CV}\int_{t}^{t+\Delta t} \frac{1}{\alpha}\frac{\partial T}{\partial t}dt\,dV \tag{A.4}$$

In Equation (A.4), the order of integration and differentiation in the term on the right-hand side has been changed to illustrate its physical meaning.

The surface integration must be carried out over the bounding surface A of the control volume. The physical interpretation of $\mathbf{n}\cdot\mathbf{a}$ is the component of the vector $\mathbf{a}$ in the direction of the outward unit vector $\mathbf{n}$ normal to infinitesimal surface element dA. The two-dimensional example of a triangular control volume is shown in Figure A.1.

Note that the bounding surface or control surface of each control volume is a closed contour formed by means of a series of finite-sized straight line elements, the area of which is denoted by ΔA. In the three-dimensional model, the control volume model would be bounded by triangular prism elements. Note that A is the area of the entire control surface in Equation (A.4) and dA indicates an infinitesimal surface element. The

Geothermal Energy: Sustainable Heating and Cooling Using the Ground, First Edition.
Marc A. Rosen and Seama Koohi-Fayegh.
© 2017 John Wiley & Sons, Ltd. Published 2017 by John Wiley & Sons, Ltd.

Figure A.1 Typical triangular control volume. In this figure, $\mathbf{n_i}$ and ΔA are the normal vector and surface element area, respectively.

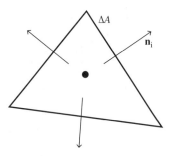

area integrations are carried out over all line segments (two-dimensional) or surface elements (three-dimensional), so they can be written as:

$$\sum_{i}^{all\ surfaces} \int_{t}^{t+\Delta t} \int_{\Delta A_i} \mathbf{n_i} \cdot \text{grad } T\, dA\, dt = \int_{CV} \int_{t}^{t+\Delta t} \frac{1}{\alpha} \frac{\partial T}{\partial t} dt\, dV \tag{A.5}$$

To evaluate the control surface integrations, expressions for grad T as well as geometric quantities, $\mathbf{n_i}$ and ΔA_i are needed. The outward normal vector $\mathbf{n_i}$ and surface element area ΔA_i can be calculated using simple trigonometry and vector algebra from the vertex coordinates of the unstructured grid (Versteeg and Malalasekera 2007). The area integration for each of the surface elements in Equation (A.5) is approximated by the dot product of the outward unit normal vector $\mathbf{n_i}$ and the diffusive flux vector (grad T) for the control surface element ΔA_i. The latter can be approximated easily using the central differencing method along line *PA*. Thus,

$$\int_{\Delta A_i} \mathbf{n_i} \cdot \text{grad } T\, dA \cong \mathbf{n_i} \cdot \text{grad } T\, \Delta A_i \cong \left(\frac{T_A - T_P}{\Delta \xi} \right) \Delta A_i \tag{A.6}$$

where $\Delta \zeta$ is the distance between the centroids A and P of two neighbor grids with a common surface ΔA_i. Note that the integration in Equation (A.6) should be carried out for all surfaces surrounding a node.

If the temperature at a node is assumed to prevail over the whole control volume, the right-hand side of Equation (A.5) can be written using first-order temporal discretization as:

$$\int_{CV} \int_{t}^{t+\Delta t} \frac{1}{a} \frac{\partial T}{\partial t} dt\, dV = \frac{1}{\alpha} \frac{T_P - T_P^0}{\Delta t} \Delta V \tag{A.7}$$

Substituting Equation (A.6) and Equation (A.7) into Equation (A.5) yields:

$$\int_{t}^{t+\Delta t} \left[\sum_{i}^{all\ surfaces} \left(\frac{T_{nb} - T_P}{\Delta \xi} \right) \Delta A_i \right] dt = \frac{1}{\alpha} \frac{T_P - T_P^0}{\Delta t} \Delta V \tag{A.8}$$

where nb is the node number of the adjacent cell. To evaluate the right-hand side of this equation we need to make an assumption about the variation of T_P and T_{nb} with time. We could use temperatures at time t or at time $t+\Delta t$ to calculate the time integral or, alternatively, a combination of temperatures at time t and $t+\Delta t$. This approach may be generalized by means of a weighting parameter θ between 0 and 1. Then, we can write the integral I_T of temperature T_P with respect to time as:

$$I_T = \int_{t}^{t+\Delta t} T_P dt = [\theta T_P + (1 - \theta) T_P^0] \Delta t \tag{A.9}$$

Using this formula for T_{nb} in Equation (A.8), and dividing by Δt throughout, we have:

$$\sum_i^{\text{all surfaces}} \left[\theta \left(\frac{T_{nb} - T_P}{\Delta \xi} \right) + (1 - \theta) \left(\frac{T_{nb}^0 - T_P^0}{\Delta \xi} \right) \right] \Delta A_i = \frac{1}{\alpha} \frac{T_P - T_P^0}{\Delta t} \Delta V \quad \text{(A.10)}$$

which may be rearranged to give:

$$\left(\frac{1}{\alpha} \frac{\Delta V}{\Delta t} + \theta \sum_i^{\text{all surfaces}} \frac{\Delta A_i}{\Delta \xi} \right) T_P = \sum_i^{\text{all surfaces}} \frac{\Delta A_i}{\Delta \xi} [\theta \, T_{nb} + (1 - \theta) T_{nb}^0] +$$

$$\left[\frac{1}{\alpha} \frac{\Delta V}{\Delta t} - (1 - \theta) \sum_i^{\text{all surfaces}} \frac{\Delta A_i}{\Delta \xi} \right] T_P^0 \quad \text{(A.11)}$$

Now, we identify the coefficients of T_{nb} and write Equation (A.11) in the familiar standard form:

$$a_P T_P = \sum_i^{\text{all surfaces}} a_{nb} [\theta \, T_{nb} + (1 - \theta) T_{nb}^0] + \left[a_P^0 - (1 - \theta) \sum_i^{\text{all surfaces}} \frac{\Delta A_i}{\Delta \xi} \right] T_P^0 \quad \text{(A.12)}$$

where

$$a_P = \theta \sum_i^{\text{all surfaces}} a_{nb} + a_P^0 \quad \text{(A.13)}$$

and

$$a_P^0 = \frac{1}{\alpha} \frac{\Delta V}{\Delta t} \quad \text{(A.14)}$$

$$a_{nb} = \frac{\Delta A_{nb}}{\Delta \xi}$$

The exact form of the final discretized equation depends on the value of θ. In the current model, the fully implicit formulation ($\theta=1$) is used. Therefore, Equation (A.12) reduces to the following form:

$$a_P T_P = \sum_i^{\text{all surfaces}} a_{nb} T_{nb} + a_P^0 T_P^0 \quad \text{(A.15)}$$

where

$$a_P = \sum_i^{\text{all surfaces}} a_{nb} + a_P^0 \quad \text{(A.16)}$$

and the constants a_P^0 and a_{nb} are introduced in Equation (A.14).

The implicit equation can be solved iteratively at each time before moving to the next time step. The advantage of the fully implicit scheme is that it is unconditionally stable with respect to time step size.

Reference

Versteeg, H.K. and Malalasekera, W. (2007) *An Introduction to Computational Fluid Dynamics. The Finite Volume Method*, 2nd edn, Prentice Hall, Harlow.

B

Sensitivity Analyses

B.1 Parameters Affecting Thermal Interactions between Multiple Boreholes

In this appendix, the effect of varying parameters such as time (t), distance between the boreholes (D_b), and heat flux from the borehole wall to the ground (q'') on the ground temperature surrounding multiple boreholes is examined using a numerical solution and compared with results of the line-source analytical solution. Note that instead of superposing one-dimensional solutions to account for circumferential heat transfer effects, the circumferential heat transfer effects are taken into account in the two-dimensional numerical solution. However, the two-dimensional numerical solution does not take into account the axial heat transfer effects in the ground. The borehole is assumed to have a constant and steady heat flow. Since the objective here is to study parameters that affect the temperature variations, a smaller heat flow rate per unit length of the borehole wall is selected (3.14 W/m along the borehole length) to reduce the size of the solution domain and computation time. The solution domain that is used for this analysis is shown in Figure B.1, and thermal properties of the ground and geometric characteristics of the boreholes are listed in Table B.1. Note that the properties of the ground are approximate values for clay soil with no water content. An initial temperature of 288 K, which is the undisturbed ground temperature, is assumed for the entire borefield. At the outer edge of the domain, a constant farfield temperature condition equal to the initial temperature (288 K) is applied. To simplify the current model, a constant heat flux of 10 W/m² on the borehole wall is assumed. In addition, to account for the transient term in Equation (A.1), the time is subdivided into 4200 time steps of 3600 s, which equals a time period of 6 months.

Figure B.2 shows the growth of the affected area in the ground with time. It is seen that the temperature gradient of the ground near the boreholes tends to decrease as time increases. Furthermore, it is noticed that, for the current analysis and assumptions for borehole distance and heat flow rate at the borehole wall, the effects of thermal interaction in terms of temperature rise are noticeable after 1 week of heat delivery to the ground. However, the temperature increase in the ground between the two boreholes due to thermal interaction does not exceed 1 K before 1 month of constant heat input.

Figure B.3 shows the effect of borehole distance (D_b) on the thermal interaction between two boreholes for analytical and numerical solutions. It is seen that, similar to the case of a 2 m distance between the two boreholes, the analytical method exhibits higher ground temperatures at the borehole wall compared with the numerical one for

Geothermal Energy: Sustainable Heating and Cooling Using the Ground, First Edition.
Marc A. Rosen and Seama Koohi-Fayegh.
© 2017 John Wiley & Sons, Ltd. Published 2017 by John Wiley & Sons, Ltd.

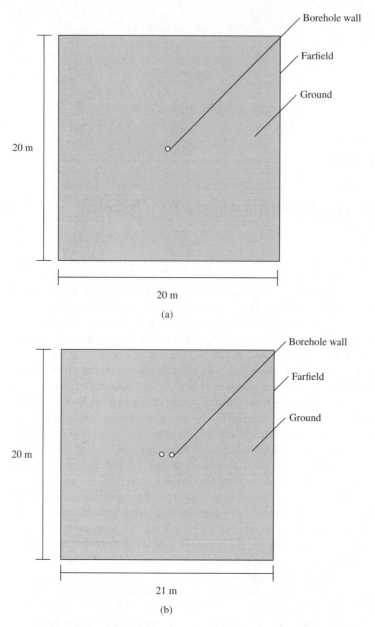

Figure B.1 Solution domain representing the ground surrounding (a) a single and (b) two boreholes, in a two-dimensional plane.

the case of a 3 m distance. Furthermore, a greater distance between the two boreholes (3 m distance compared with 2 m) leads to a weaker interaction between the two boreholes; the temperature of the ground between the two boreholes after 6 months decreases from 289.5 K for the case of D_b=2 m to 289.1 K for the case of D_b=3 m. The same trend is noticed for the temperature at the borehole wall, which is 290.7 K for a smaller distance (D_b=2 m) and decreases to 290.5 K for a greater distance (D_b=3 m).

Table B.1 Thermal properties and geometric characteristics of the model.

Parameter	Value
Ground	
Undisturbed ground temperature	15 °C (288 K)
Ground thermal conductivity	1 W/m K
Ground specific heat capacity	1200 J/kg K
Ground density	1381 kg/m³
Borehole geometry	
Borehole radius, r_b	0.050 m
Number of boreholes	2
Borehole separation distance, D_b	2 m
Heat flow rate per unit length, q'	3.14 W/m

Source: Adapted from Shonder and Beck (1999), Incropera and DeWitt (2000), Hepbasli *et al.* (2003), and Gao *et al.* (2008).

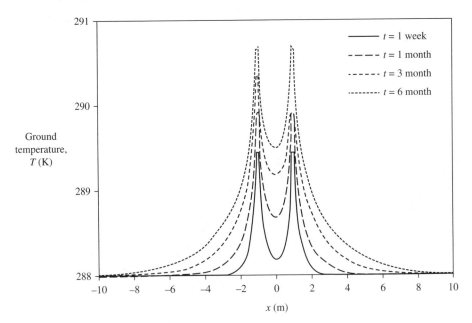

Figure B.2 Temperature of ground surrounding multiple boreholes at several values at time *t* at $y=0$ m.

Figure B.4 shows the temperature of the ground surrounding the two boreholes installed at different separation distances. Note that the closer the two boreholes are installed, the stronger is the thermal interaction between them (the temperature between them reaches 290.2 K for $D_b=1$ m compared with 289.1 K for $D_b=3$ m) and the higher becomes the temperature of ground at the borehole wall (291 K for $D_b=1$ m

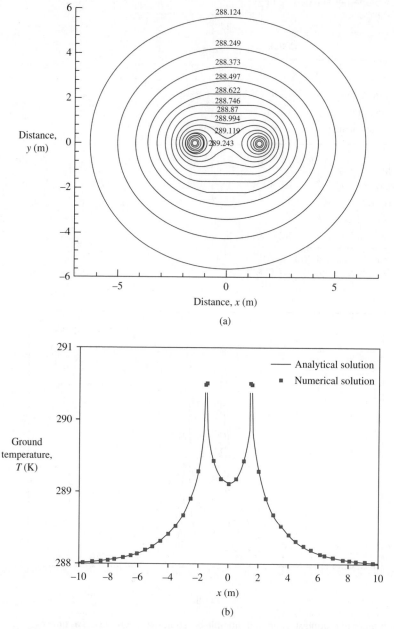

Figure B.3 Ground temperature around multiple boreholes at $t=6$ months and $D_b=3$ m. (a) Temperature contours (K) of the analytical solution. (b) Comparison of the analytical and numerical solutions at $y=0$ m.

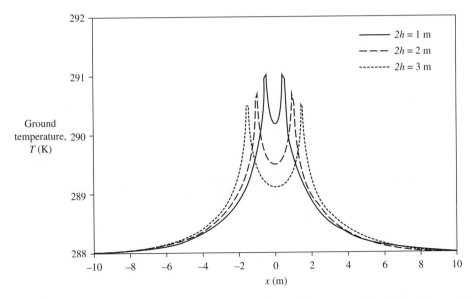

Figure B.4 Temperature of ground surrounding multiple boreholes at several borehole distances at $y=0$ m.

compared with 290.5 K for $D_b=3$ m). For a specific heat flux from the borehole wall, a borehole separation distance can be calculated in order for the temperature of the ground to stay below a desired limit. It is also observed that, for a specified heat flux on the borehole wall (10 W/m²), a greater distance between the two boreholes results in a slightly larger region in the ground experiencing a temperature excess of more than 0.05 K (7.6 m for $D_b=1$ m compared with 8.1 m for $D_b=3$ m). However, since the interaction effect is smaller for higher borehole distances, moving away from each borehole towards farfield, there is a larger temperature gradient for boreholes with a larger spacing, that is, the temperature excess in the ground disappears at a shorter distance from the borehole. This can also be due to less pronounced temperature rise effects from each borehole reaching the outer ground surrounding the other borehole.

The effect of heat flux at the borehole wall on the thermal interaction between the two boreholes for both analytical and numerical solutions is shown in Figure B.5. It is seen that a larger heat flux results in a significant increase in the temperature of the ground at the borehole wall (301.5 K for $q''=50$ W/m²) and also in the region far from the borehole wall. This results in a stronger thermal interaction between the two boreholes as well. As can be seen in Figure B.5, the temperature of the ground between the two boreholes increases from 289.5 K in the case of $q''=10$ W/m² to 295.7 K in the case of $q''=50$ W/m². It is noticed that for a higher heat flux ($q''=50$ W/m² compared with $q''=10$ W/m²), the results of the numerical solution still match the analytical results well.

Figure B.6 shows the effect of heat flux at the borehole wall on the temperature response of the ground surrounding the boreholes. It is seen that a higher heat flux results in a higher temperature at the borehole wall and also a greater area around the borehole experiencing a temperature excess. A comparison of the affecting parameters on the thermal interaction between the two boreholes reveals that varying the heat

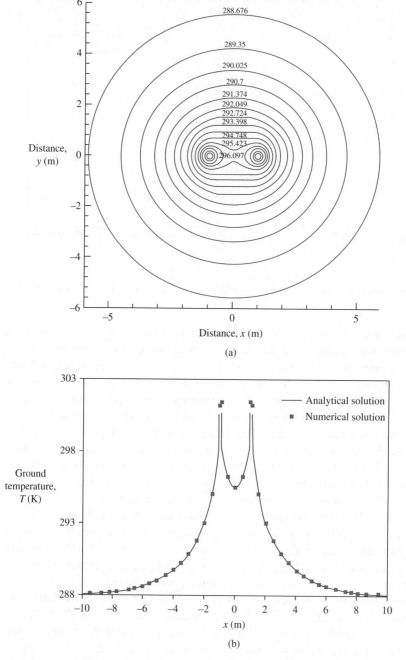

Figure B.5 Ground temperature around multiple boreholes at $q''=50\,W/m^2$ and $t=6$ months. (a) Temperature contours (K) of the analytical solution. (b) Comparison of the analytical and numerical solutions at $y=0$ m.

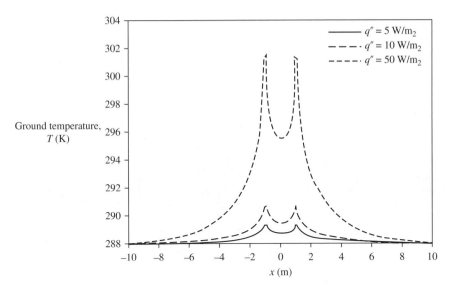

Figure B.6 Temperature of ground surrounding multiple boreholes for various values of heat flux (q'') at the borehole wall.

flux at the borehole wall has a greater impact on the thermal interaction between the boreholes than varying borehole distances with the same ratio.

Figure B.7 shows the effect of varying heat flux from the borehole walls. It is seen that, even with varying amounts of heat input in the ground, the temperature of the ground surrounding the borehole is affected almost symmetrically in the form of circular contours at significant distances (greater than about 3 m). A comparison of the numerical and analytical results (Figure B.7b) shows that except for the ground at the borehole wall, similar to the previous cases, the results match. It is seen in this figure that if the sum of heat fluxes from the two boreholes is the same, the temperatures of the ground surrounding the boreholes match beyond a distance of about 6 m for the case of $D_b=2$ m.

It is worth mentioning that the methods used for calculating the temperature profiles in the ground surrounding two boreholes can be applied to two systems of borehole heat exchangers as well. For example, if an area of 40 m × 40 m × 200 m in the ground is occupied for one system of borehole heat exchangers, the ratio of system depth to its radial size is large enough to be treated as one cylinder or line source of heat when system interactions and temperature excess around a system in larger distances are to be accounted for. Therefore, a parametric study on two interacting boreholes can determine the results for two interacting systems of boreholes.

B.2 Validation of the Two-Dimensional Numerical Solution with a Three-Dimensional Solution

One of the drawbacks of the two-dimensional numerical solution presented in Section B.1 is that it neglects axial heat transfer effects, which exist at depths near the top and bottom of the boreholes. To examine the inaccuracy associated with this simplification in the evaluation of the temperature response of the ground surrounding the

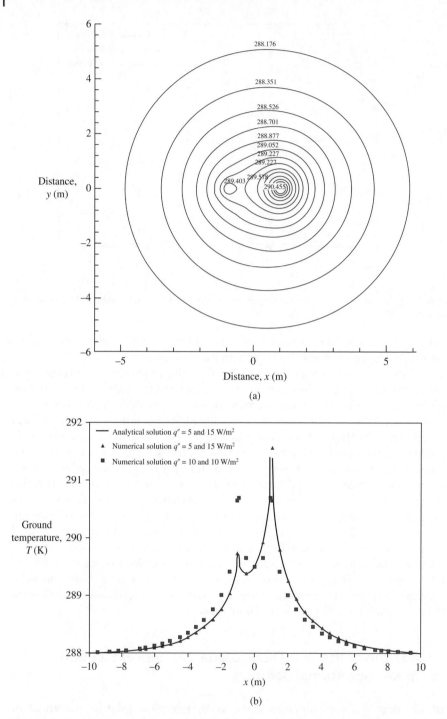

Figure B.7 Ground temperature around multiple boreholes at $q''=5$ and 15 W/m^2 and $t=6$ months. (a) Temperature contours (K) of the analytical solution. (b) Comparison of the analytical and numerical solutions at $y=0$ m.

boreholes, a three-dimensional solution domain representing the ground surrounding the boreholes is considered in this section and the manner in which borehole axial effects cause variations in the temperature profile is examined. Note that the heat flux from the borehole wall to the ground is assumed to be constant along the borehole length at $q''=10\,\text{W/m}^2$. Since the objective of the analysis in this section is only to compare the two- and three-dimensional solution domains, the selected heat flow rate is a relatively lower value compared with the design range ($30-60\,\text{W/m}$) to reduce the solution domain size and computation time.

The heat transfer symmetry about the two vertical planes shown in Figure B.8a is utilized and, since geometric symmetry along the borehole length exists, it is assumed that

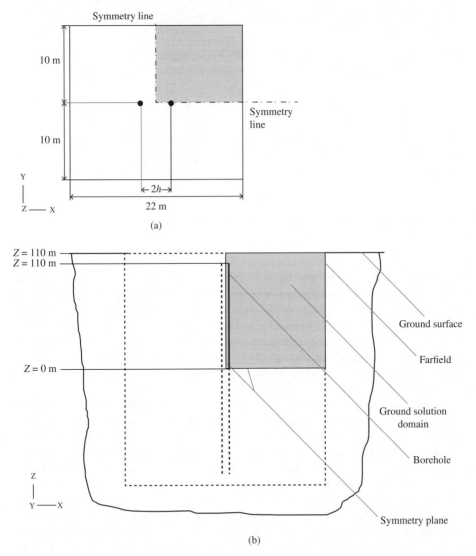

Figure B.8 Two-dimensional view of the solution domain. (a) Horizontal cross sections (*xy*) at the borehole mid-length (*z*=0 m) and (b) vertical cross section (*xz*).

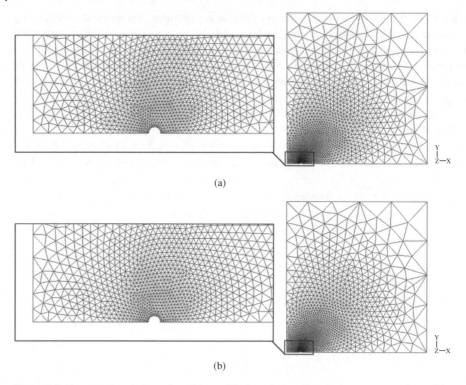

(a)

(b)

Figure B.9 Computational triangular grids used in the solution domain in *xy* cross section at (a) *z*<100 m and (b) *z*>100 m.

the heat transfer along the borehole is symmetrical about a horizontal plane passing through the borehole mid-length (Figure B.8b). Therefore, only one-eighth of the borehole field is modeled and the solution domain (ground) is enclosed by the farfield, the ground surface, and three symmetry planes. In Figure B.8b, the gray area is the solution domain, the results of which can be replicated to the other areas drawn with dashed lines due to their symmetry. Since the temperature gradient in the domain between the borehole wall and the farfield changes gradually from large to small, the size of the mesh cells is chosen based on this gradual change to reduce computer memory and computational time (Figure B.9).

An initial temperature of 288 K, which is the undisturbed ground temperature, is assumed for the entire borefield. At the outer edge of the domain, a constant farfield temperature condition equal to the initial temperature (288 K) is applied. The conditions at the symmetry planes are set for zero heat flux. To simplify the current model, a constant heat flux of 10 W/m^2 on the borehole wall is assumed since, in order to study the thermal interaction between multiple boreholes, their inner dynamic heat exchange process can be of second priority compared with the heat dissipation in the ground surrounding them. In the three-dimensional analysis, borehole lengths of 200 m (Figure B.8b) and an adiabatic heat transfer condition for the ground surface is assumed. Using these boundary conditions, and the physical and geometric properties listed in Table B.2, the integral form of conduction equation for the conservation of energy [Equation (A.1)] is solved using a control volume method in ANSYS FLUENT software. To account for the

Table B.2 Thermal properties and geometric characteristics of the model.

Parameter	Value
Ground	
Undisturbed ground temperature	15 °C (288 K)
Ground thermal conductivity	1 W/m K
Ground specific heat capacity	1200 J/kg K
Ground density	1381 kg/m^3
Borehole geometry	
Total borehole length, H	200 m (three-dimensional case)
Borehole radius, r_b	0.050 m
Number of boreholes	2
Borehole distance, D_b	2 m
Heat flow rate per unit length, q'	3.14 W/m

Source: Adapted from Shonder and Beck (1999), Incropera and DeWitt (2000), Hepbasli *et al.* (2003), and Gao *et al.* (2008).

transient term in Equation (A.1), the time is subdivided into 4200 time steps of 3600 s, which equals a time period of 6 months.

Figure B.10 compares the temperature profiles in the ground surrounding multiple boreholes evaluated by the two-dimensional solution with the ground temperature profiles evaluated by the three-dimensional method at distances in the middle of the borehole length ($z=0$) and 4 m away from the bottom of the borehole ($z=96$ m). It is seen that the temperature values calculated by the two-dimensional method agree well with the results from the three-dimensional method for about 96% of the borehole length (less than 1% error in ground temperature). Therefore, it is concluded that the two-dimensional method, having a comparatively lower computational time, is valid for calculating the temperature response of the ground around 96% of the borehole length.

The temperature response of the ground surrounding multiple boreholes evaluated by the three-dimensional solution at various borehole depths is illustrated in Figure B.11. It is shown that the temperature rise in the ground surrounding the borehole decreases at the very end of borehole length where axial heat transfer effects come into play. The maximum amount of temperature rise due to thermal interaction of multiple boreholes in a 6-month period of heat transfer from the borehole to the ground occurs in the middle of the borehole length ($z=0$). Therefore, with the objective of limiting the operations and sizes of the boreholes in order to prevent their thermal interaction, the middle length of the boreholes is the critical area. This is true when the heat flux from the borehole wall to the surrounding ground does not change significantly along the borehole length and the assumption of constant heat flux from the borehole wall is valid.

In order to examine the validity of the two-dimensional results for a higher heat flux from the borehole wall, Figure B.12 shows the results of a three-dimensional analysis for $q''=20$ W/m^2. It is seen that, although a larger temperature rise around the boreholes and a larger thermal interaction between them exist for this case, the temperature profile

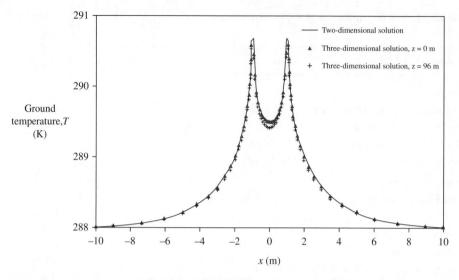

Figure B.10 Ground temperature around multiple boreholes at $q''=10\,\text{W/m}^2$ and $t=6$ months, and comparison of the two- and three-dimensional solutions at various borehole depths.

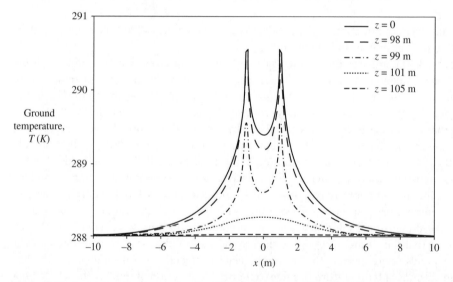

Figure B.11 Ground temperature around multiple boreholes at $q''=10\,\text{W/m}^2$ and $t=6$ months, and temperature response of the ground at various borehole depths for the three-dimensional analysis.

around the boreholes does not change in the middle of the borehole length ($z=0$ m) and until about 4 m away from the top and bottom of the boreholes ($z=96$ m). Therefore, similar to the previous case ($q''=10\,\text{W/m}^2$), axial heat transfer effects are negligible for about 96% of the borehole length and the two-dimensional analysis can be applied for temperature evaluation of the ground surrounding the boreholes. It is concluded that the length of the borehole for which axial heat transfer effects are negligible and two-dimensional analysis can be applied does not vary with the change in the amount of heat flux from the borehole wall to the ground.

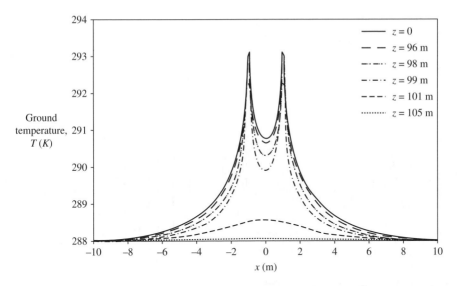

Figure B.12 Ground temperature around multiple boreholes at $q''=20$ W/m^2 and $t=6$ months, and temperature response of the ground at various borehole depths for the three-dimensional analysis.

B.3   Heat Flux Variation along Borehole Length

A limitation in the models presented in the previous section is the assumption of uniform heat to the ground along the borehole length, when the borehole is assumed as a line source of heat. In reality, however, the temperature and heat flux distributions on the borehole wall can only be decided by accounting for the heat exchange process between the tubes in the borehole and the borehole wall. A variable heat flux (VHF) along the borehole is calculated by defining the temperature profiles of the fluid running through the tubes in the borehole.

It should be noted that the current section focuses only on the variation of heating strength along the borehole length. Since only the existence of such a variation is intended to be discussed, the current section does not provide typical values for the borehole spacing and the heat flux on the borehole wall, and lower values are chosen in order to keep the solution domain size smaller in the numerical solution. Note also that the temperature of the ground at the borehole wall which is used in coupling the model inside the borehole with the one outside the borehole in the current problem is assumed to be constant throughout the entire operation time (see Section 8.1.3.1 for more details on the model). Therefore, the current solution is only valid for low temperature variations in the ground surrounding the boreholes, which only occurs by assuming lower heat flux values on the borehole wall. Modifying the current problem to one with typical industrial values for ground-source heat pump systems requires the ground temperature to be assumed variable and is the subject of Section 8.2. The discussions of this section are intended to highlight the importance of choosing a variable borehole wall temperature when coupling the models for inside and outside the borehole by using a method that updates this value at every time step and one that does not.

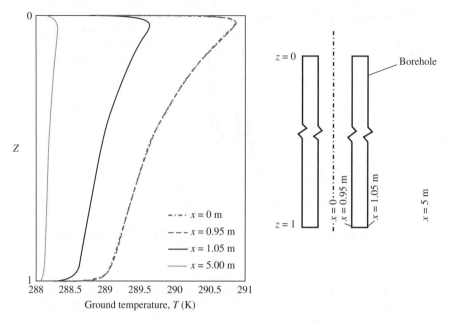

Figure B.13 Ground temperature around multiple boreholes in the *xz* plane in *t*=6 months, at various distances from borehole wall for the variable heat flux model.

A three-dimensional model of transient conduction of heat in the ground surrounding multiple ground heat exchangers is presented in this section. A domain consisting of two vertical borehole heat exchangers having a distance of D_b from each other is considered.

The temperature responses of the ground surrounding multiple boreholes evaluated by the VHF model at various borehole depths are compared in Figure B.13 and Figure B.14a. It is shown in Figure B.13 that the maximum temperature rise due to thermal interaction of multiple boreholes in a 6-month period of heat transfer from the borehole into the ground occurs at the top 3% heating length of the borehole and it decreases along the borehole length as the heat flux from the borehole wall to the ground decreases. Therefore, with the objective of limiting the operations and sizes of the boreholes in order to prevent their thermal interaction, the top length of the boreholes (about 3% total length) is the critical area. Also, as expected the maximum temperature rise in the ground occurs at the borehole wall (*x*=0.95 m and *x*=1.05 m). Since the current analysis does not use typical conditions (e.g., values for borehole spacing, heat flux on the borehole wall), a minimum value of spacing is not suggested in this analysis. An extension of the current analysis to typical industrial values may require the assumptions of constant borehole wall temperature and constant ground surface temperature made in the current model to be modified to be variable (see Section 8.2 for such an analysis). In such a case, using the current solution method, it is possible to determine a minimum value of spacing or maximum amount of heat input to the ground to avoid thermal interactions between boreholes under typical conditions.

It is shown in Figure B.14a that the thermal interaction between the boreholes is at a minimum at the bottom of the borehole (*z*=−99.9 m) where the heat flux to the ground is lowest. This is not true for the case of constant heat flux from the borehole wall to the surrounding ground along the borehole length (Figure B.14b). It is seen in Figure B.14b

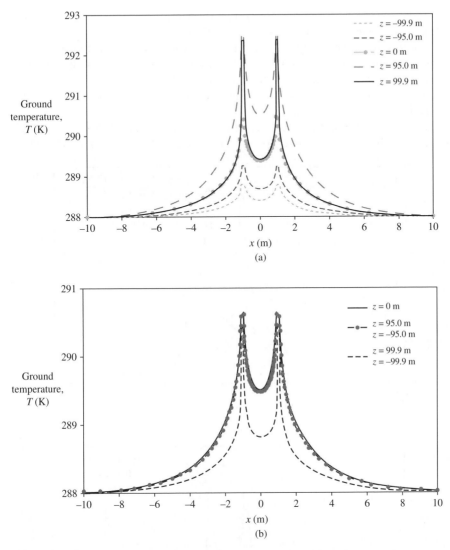

Figure B.14 Ground temperature around multiple boreholes in $t=6$ months, at various borehole depths for (a) the variable heat flux model and (b) the constant heat flux model.

that the greatest thermal interaction occurs at top of the borehole, but remains at its maximum along the borehole length. For this case, the critical length of the borehole would be almost 95% of the borehole length. However, as discussed earlier, the case of constant heat flux is only a simplification of the VHF problem and does not present the problem as accurately as the VHF problem. Note that in order to compare the results from the constant heat flux model with the results of the VHF model, an equivalent inlet temperature ($T'_f = 290.6$ K) for the VHF model, resulting in the same total heat conduction in the ground, is assumed. The total heat flow rate is calculated by integrating the heat flow rate along the borehole.

Another notable characteristic of Figure B.14 is the decrease in the thermal interaction at lengths of $z=99.9$ m when one moves from $z=95$ m towards the top end of the borehole. Specifically, for the case of VHF, there is a higher heat flux as one moves towards the top end and one expects greater thermal interactions. In both cases, the temperature rise in the ground surrounding the borehole declines at the very end of the borehole length, and this is likely due to axial heat transfer effects which become notable only at the very ends of the borehole lengths.

The results of the VHF and constant heat flux models are compared in Figure B.15. It is seen in Figure B.15a that the assumption of constant heat flux on the borehole wall

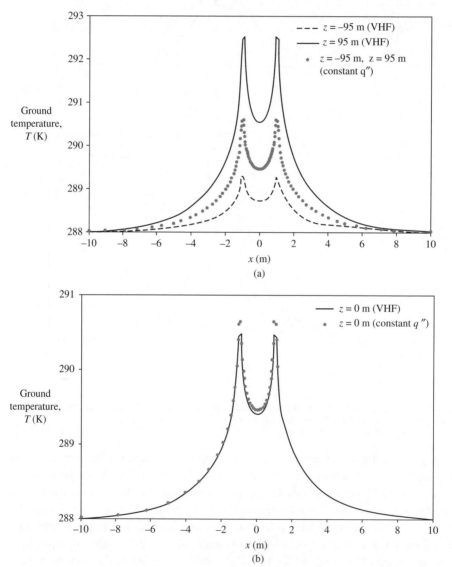

Figure B.15 Comparison of ground temperature around multiple boreholes at $t=6$ months for variable heat flux (VHF) and constant heat flux models at (a) $z=95$ m and $z=-95$ m and (b) $z=0$ m.

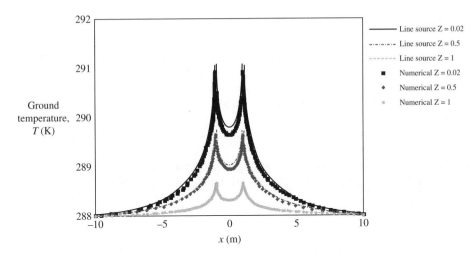

Figure B.16 Ground temperature around multiple boreholes in $t=6$ months for line-source and numerical models at various borehole depths.

introduces numerous inaccuracies especially when dealing with the temperature rises in the ground at the very top and bottom of the borehole. Figure B.15b shows that, by using the VHF method, the heat flux on the borehole is spread along the borehole in a way that the heat flux in the middle depth remains similar to its average. It can be concluded that using the constant heat flux method is only valid for the middle length of the boreholes and moving further to the top or bottom of the borehole, the evaluated temperature rises become increasingly inaccurate. Quasi-three-dimensional models reveal drawbacks of two-dimensional models and are thus preferred for the design and analysis of ground heat exchangers, as they provide more accurate information for performance simulation, analysis, and design.

Note that the effect of temperature rise due to one borehole on the other is neglected by applying the superposition method. This effect is examined for a two-dimensional numerical domain in Figure 7.3b and Figure 7.4 by comparing the results of the numerical solution with analytical results of line-source theory where the superposition method is used to account for the temperature rise in the ground surrounding multiple boreholes. It is shown that these effects are minor in comparison with the order of the temperature rise in the ground due to the individual performances of the boreholes. Since the objective in the current section is to examine at what depths the thermal interaction among boreholes creates a critical temperature rise, the focus is mostly on introducing a heat flow rate profile along the borehole length which can be coupled to the numerical or line-source model outside the borehole to show the effect of VHF along borehole length on temperature rise in the soil. In Figure B.16, the two methods are compared and it is shown that the temperature rise in the ground caused by both methods is very close. Therefore, it is concluded that the analytical method presented in Section 8.1.3.1 can provide as accurate results as a numerical method.

References

Gao, J., Zhang, X., Liu, J. *et al.* (2008) Numerical and experimental assessment of thermal performance of vertical energy piles: An application. *Applied Energy*, **85**, 901–910.

Hepbasli, A., Akdemir, O. and Hancioglu, E. (2003) Experimental study of a closed loop vertical ground source heat pump system. *Energy Conversion and Management*, **44**, 527–548.

Incropera, F.P. and DeWitt, D.P. (2000) *Introduction to Heat Transfer*, 3rd edn, John Wiley & Sons, Inc., New York.

Shonder, J.A. and Beck, J.V. (1999) Determining effective soil formation thermal properties from field data using a parameter estimation technique. *ASHRAE Transactions*, **105**, 458–466.

Index

Geothermal Energy: Sustainable Heating and Cooling Using the Ground, First Edition.
Marc A. Rosen and Seama Koohi-Fayegh.
© 2017 John Wiley & Sons, Ltd. Published 2017 by John Wiley & Sons, Ltd.